Springer Series in Statistics

Springer Series in Statistics

Gavin J.S. Ross

Nonlinear Estimation

With 57 Figures

Springer-Verlag
New York Berlin Heidelberg
London Paris Tokyo Hong Kong

Gavin J.S. Ross
Statistics Department
Institute of Arable Crops Research
Rothamsted Experimental Station
Harpenden
Herts AL5 2JQ
England

Mathematical Subject Classifications: 62-xx, 62Kxx, 62J02, 65K10, 65D10, 62F25

Library of Congress Cataloging-in-Publication Data
Ross, Gavin J.S.
    Nonlinear estimation/Gavin J.S. Ross.
    p. cm.—(Springer series in statistics)
    ISBN-13:978-1-4612-8001-9
    1. Estimation theory. 2. Nonlinear theories. I. Title.
    II. Series.
    QA276.8.R67   1990
    519.5'44—dc20   90-32797

Printed on acid-free paper.

Typeset by Asco Trade Typesetting Ltd., Hong Kong.

9 8 7 6 5 4 3 2 1

ISBN-13:978-1-4612-8001-9   e-ISBN-13:978-1-4612-3412-8
DOI: 10.1007/978-1-4612-3412-8

Contents

Introduction

This work describes the theory of nonlinear estimation in the context of fitting models to data. The research is based on computing experience over the years, arising from the development of the Maximum Likelihood Program (MLP), which fits a wide range of nonlinear models. Users of this program have contributed greatly by reporting cases of data that the program was unable to fit, and by suggesting new models or analyses that could aid the interpretation of the results.

The general questions to be discussed are:

(1) What models are appropriate for interpreting data?
(2) How may different models be compared?
(3) Why are some data sets more easily fitted than others?
(4) How can estimates be obtained speedily and accurately?
(5) What does it mean if parameter estimates are not unique or are non-existent?
(6) How do we best describe the uncertainty in the estimates of parameters and other quantities?
(7) How does each item of data affect the estimates of parameters and other quantities?

The models to be discussed will mainly be curves, surfaces, frequency distributions, discrete probability models, and differential equation models, with appropriate assumptions about error distributions. The unifying framework is maximum likelihood theory, both as a basis for estimation and for comparison of models, although alternative methods of estimation will be discussed. The treatment relies on statistical plausibility, computing experience, and geometrical reasoning rather than on pure algebraic manipulation.

Some results cannot be proved rigorously; the exceptions are usually unimportant, we hope!

A basic idea is to select reparameterizations with good properties, noting that this does not change the relationship between the data values and their corresponding fitted values. Estimation theory normally uses the basic parameters of the model; asymptotic results based on series expansions pertain to large samples, or small error variances. Transformed parameters may allow the model to be specified in terms much closer to linearity for small samples. Reparameterization does not invalidate traditional asymptotic theory, but it does introduce an extra dimension that may be exploited for a wide range of ordinary model-fitting problems.

Chapter 1 describes the scope of this work and the statistical theory on which it is based. Chapter 2 describes how parameters may be transformed, while Chapter 3 discusses problems of inference and design. Chapter 4 describes a range of graphical methods useful in the study of nonlinear models. Chapter 5 describes computational aspects of nonlinear modeling. Some applications are discussed in Chapter 6, while in Chapter 7 the standard models in MLP are described in detail. A glossary of unfamiliar terms is given in the Appendix.

My thanks are due to John Gower for his encouragement, to Valerie Payne for typing the manuscript, to the staff of Springer-Verlag for their cooperation, and to the many users of nonlinear models whose problems have stimulated this work.

CHAPTER 1

Models, Parameters, and Estimation

1.1. The Models To Be Considered

The class of models to be discussed is a generalization of the multiple regression model. Observations $\mathbf{y} = y_1, \ldots, y_n$ are assumed to be random samples from a specified distribution $f(\mathbf{y}|\boldsymbol{\theta})$ where $\boldsymbol{\theta} = \theta_1, \ldots, \theta_p$ is a vector of parameters. The observations may be raw data or functions of raw data such as logarithms or proportions. The mean of the distribution, $\boldsymbol{\mu} = E(\mathbf{y})$, also known as the vector of *expectations*, is often a function of one or more independent variables, x_1, x_2, etc., and the parameters. The distribution about the mean, referred to as the *error distribution* or *random part* of the model, may involve further parameters representing variances, correlations, or weight functions.

In practice, the most widely used distributions are the normal, lognormal, and gamma for continuous variables, and the Poisson, binomial, and multinomial for discrete variables. Observations may be independent or correlated, and given *a priori* equal or unequal weights which may depend on expectations.

Some examples will now be given.

1.1.1. Curves

Data $\mathbf{y}$ are continuous or discrete variables and depend on a single independent variable $\mathbf{x}$. Thus, for example,

$$E(\mathbf{y}|\mathbf{x}, \boldsymbol{\theta}) = \theta_1 + \theta_2 \exp(-\theta_3 \mathbf{x}) \qquad (1.1.1)$$

is the exponential curve with unknown asymptote θ_1. Polynomials, and trigonometric curves such as

$$E(\mathbf{y}) = \theta_1 + \theta_2 \sin(\mathbf{x}) + \theta_3 \cos(\mathbf{x}), \qquad (1.1.2)$$

are linear with respect to the parameters, the latter becoming nonlinear if the period is unknown. The most widely used curves (illustrated in Fig. 1.1) are:

(1) polynomials, important because of their statistical linearity;
(2) exponentials, sometimes with fixed origin or asymptote, arising as solutions of simple linear differential equations, or as useful empirical forms;
(3) hyperbolas, or ratios of polynomials, chosen as empirical forms with asymptotes, or arising from simple physical or chemical models;
(4) growth curves, such as the logistic (inverse exponential) or Gompertz (exponential of exponential), arising from differential equations for growth processes;
(5) Gaussian curves, for spectrographic data, etc.;
(6) trigonometric (Fourier) curves, for periodic phenomena;
(7) compounds of the above in which two or more underlying processes occur together; and
(8) splines, or segments of separate curves with continuity at the joins (or nodes).

The error distribution depends on the basic form of $\mathbf{y}$. For unlimited continuous data the normal distribution is often appropriate. When $\mathbf{y}$ is measured at regular intervals of time, successive observations may be correlated. Data constrained to be positive may be modeled more appropriately by the gamma or lognormal distribution, possibly with unknown origin if a constant coefficient of variation cannot be assumed. Quantal response data (proportions of a binomial sample), widely observed in biological assays and similar investigations, are fitted to curves constrained to lie between 0 and 1.

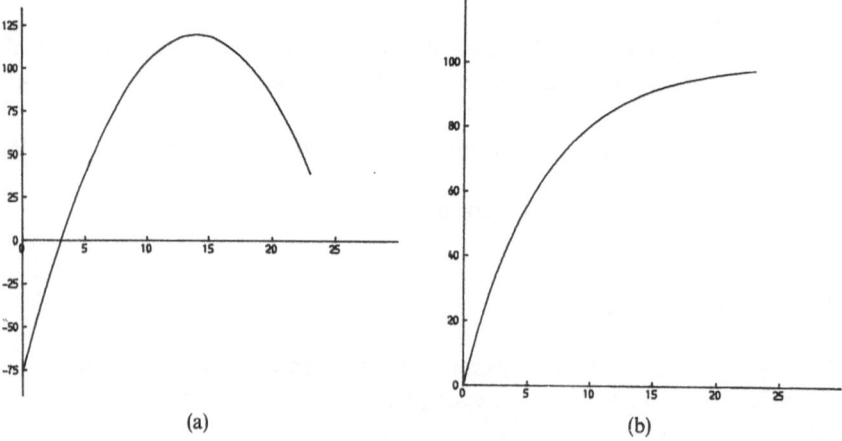

Fig. 1.1. Curves widely used for fitting to data: (a) quadratic, (b) exponential, (c) rectangular hyperbola, (d) logistic, (e) Gaussian, (f) sine, (g) exponential + straight line, (h) two straight lines.

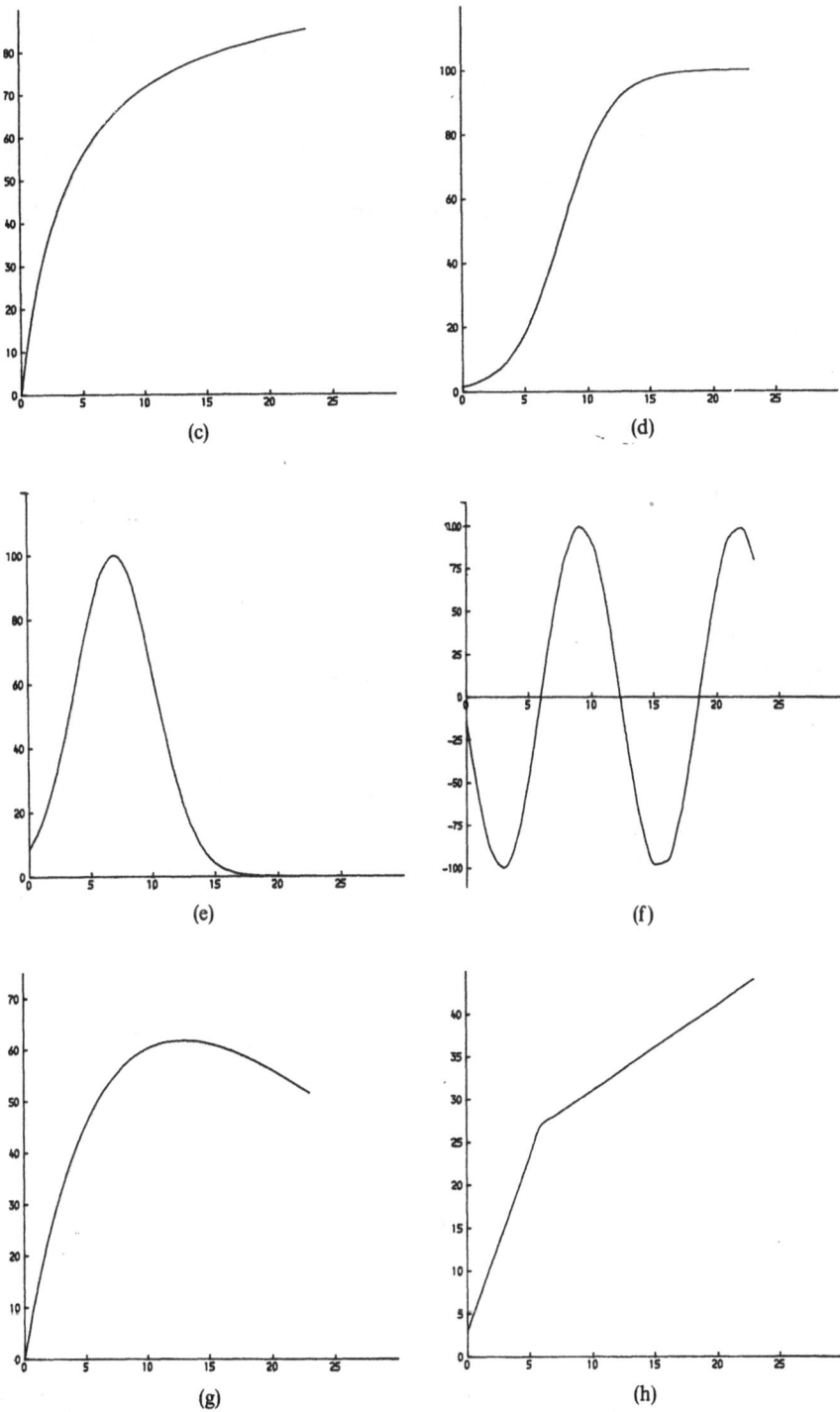

Fig. 1.1 (*cont.*)

1.1.2. Parallel Curves

When straight lines are fitted to several sets of data it is often of interest to test whether their slopes differ significantly. The simpler model with common slope but different intercepts for each line is easily fitted, producing a set of parallel regression lines. This analysis may be generalized in many ways.

When the same curve is fitted to more than one set of data it is useful to fit compound models in which some parameters are to be common to all data sets and the remainder specific to each data set. These models are especially useful when data sets are to be compared. Several formulations are possible, for example, in the case of the exponential curve (1.1.1).

(1) *Calibration*, or horizontal displacement

$$E(y_j) = \theta_1 + \theta_2 \exp(-\theta_3(x_j - \alpha_j)), \qquad (1.1.3)$$

where $\alpha_1 = 0$ and the differences between the α parameters are to be estimated.

(2) *Vertical displacement*

$$E(y_j) = \theta_{1j} + \theta_2 \exp(-\theta_3 x_j), \qquad (1.1.4)$$

in which the model is identical in each set apart from the height at any point.

(3) *Constant underlying shape*

$$E(y_j) = \theta_{1j} + \theta_{2j} \exp(-\theta_3 x_j), \qquad (1.1.5)$$

in which each set estimates its own origin and scale, given the common shape of the curve (in this case, θ_3 is the rate of approach to the asymptote). These cases are illustrated in Fig. 1.2.

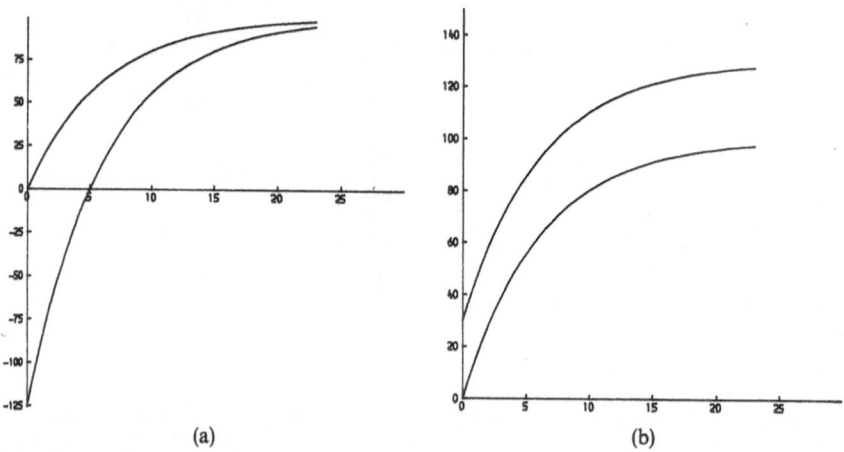

(a)   (b)

Fig. 1.2. Parallel exponential curves: (a) displaced in the x direction; (b) displaced in the y direction; (c) displaced in both the x and y directions.

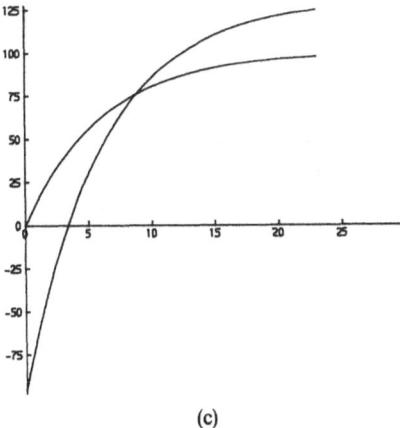

(c)

Fig. 1.2 (*cont.*)

1.1.3. Surfaces and Higher-Dimensional Relationships

With two or more independent variables, the number of possible models increases very rapidly. Surfaces may be generated as sums, ratios, or products of curves in the separate variables, or as more general expressions in the variables. There is no systematic study of surface fitting. The problems of curve fitting are compounded when extended to surfaces. Major areas of application are in contour estimation, and in fitting complex physical and chemical relationships (see Figs. 4.1 and 4.2).

1.1.4. Frequency Distributions

Data sampled from a frequency distribution, which may be of one or more discrete or continuous random variables, can be included in the class of problems to be discussed. For discrete distributions with probability function $p(\mathbf{y}|\boldsymbol{\theta})$ the log-likelihood of a sample $\mathbf{y}$ of n observations is

$$\sum_i \log p(y_i|\boldsymbol{\theta}),$$

and if the observations are classified into c different values $Y_1, \ldots, Y_c$ with n_j occurrences of Y_j, the log-likelihood is

$$\sum_j^c n_j \log p(Y_j|\boldsymbol{\theta}).$$

If the number of possible classes is infinite, the less frequent classes may be pooled. For one-dimensional distributions the probability associated with the "tails" may be computed assuming the complete distribution has probability 1.

For continuous distributions with frequency function $f(y|\theta)$ the log-likelihood is

$$\sum_i \log f(y_i|\theta).$$

With some loss of accuracy this may be approximated by classifying similar values of y_i into c classes, such as

$$Y_{j-1} < y_i \leq Y_j,$$

and if

$$p_j(\theta) = \int_{Y_{j-1}}^{Y_i} f(y|\theta)\, dy$$

the log-likelihood for n_j occurrences of class j is

$$\sum_j n_j \log p_j(\theta),$$

which is the same as for a multinomial distribution with probabilities $p_j(\theta)$.

1.1.5. Discrete Models

Many special models occur in the biological and physical sciences. For example, the population dynamics of a particular species can be modeled by assuming that given an initial age distribution at time t_0, the age distribution at time t_1 depends on the proportions dying, giving birth, or entering a new age group between times t_0 and t_1. Data observations consist of samples at particular points in time. The results of genetic experiments may be expressed in terms of the predicted frequencies of particular genotypes, modeled as a set of probabilities in a multinomial sample.

Taken to the limit by shortening the time interval to infinitesimal amounts, the population dynamics example becomes a model defined as a differential equation system, which can either be integrated explicitly to give a set of curves with algebraic form, or be integrated numerically to realize expectations corresponding to data observations sampled at discrete time intervals (see Section 6.3).

1.2. Maximum Likelihood Estimation

The asymptotic theory of maximum likelihood estimation is described in standard textbooks such as Kendall and Stuart (1951), Cramer (1946), and Cox and Hinkley (1974). The purpose of this section is to set down the relevant results below.

Given a sample y from a probability distribution $p(y|\theta)$, the expression of p as a function of parameters θ is known as the *likelihood* of θ given y. Values

$\hat{\theta}$ of θ which maximize the likelihood are known as *maximum likelihood estimates* (MLEs) of θ. In practice, it is more convenient to minimize the function $L(\theta) = -\log p(\mathbf{y}|\theta)$ which is usually termed the *log-likelihood*. There is, of course, a monotone relationship between likelihood and log-likelihood.

1.2.1. Properties of Maximum Likelihood Estimates

Maximum likelihood estimates are not necessarily unique: the likelihood function may have several local maxima, or none at all. It is a matter of taste whether local maxima should be called MLEs. Where there is no analytical maximum within the allowed range of θ, the absolute maximum is sometimes termed the MLE although the partial derivatives $\partial L/\partial\theta$ do not vanish (see Section 3.1).

Maximum likelihood estimates are *consistent* in the sense that as the sample size increases, $\hat{\theta}$ tends asymptotically to θ.

Maximum likelihood estimates are *asymptotically efficient* in the sense that as sample size increases their *dispersion matrix*, $D(\theta)$, which is the inverse of the *information matrix*,

$$I(\hat{\theta}) = E\left(\frac{\partial L(\theta)}{\partial\theta_i\,\partial\theta_j}\right)_{\theta=\hat{\theta}},$$

attains the lower bound of the Cramer–Rao inequality. This means that in large samples non-MLEs cannot have a greater variance than the MLE.

Maximum likelihood estimates are *asymptotically normally distributed* with the dispersion matrix, $D(\hat{\theta})$. But this result is not very useful in practice because in most practical cases of nonlinear modeling the sample size is not large enough for the nonnormality to be ignored (see Section 2.1.1).

Maximum likelihood estimates may be *biased* in small samples, in the sense that the expectation of $\hat{\theta}$ does not necessarily equal θ. If the parameters are transformed nonlinearly, say from θ_i to $\log(\theta_i)$, the estimation problem is unaltered but the mean of $\log(\theta_i)$ is not equal to the log of the mean of θ_i, so that even if θ is unbiased in one parametrization it must be biased in others, depending on how the parameters are defined. Therefore, if bias is a problem it may be reduced by finding an appropriate transformation (see Section 2.1.2).

1.2.2. Hypothesis Testing and Confidence Regions

The fundamental theorem of Neyman and Pearson (1928) concerns the distribution of the increase in the minimized log-likelihood when some parameters are fixed.

The minimum value of the log-likelihood, $2L(\hat{\theta})$, is now widely known as the *deviance* (Nelder and Wedderburn, 1972). Given a model with p parameters θ and q parameters ϕ, let $2L_{p+q}$ be the deviance after fitting the full set of $p + q$ parameters, and let $2L_p$ be the deviance fitting θ only, with fixed values of ϕ. Then the Neyman and Pearson theorem states that the *likelihood ratio criterion*, $\gamma = 2(L_p - L_{p+q})$, is asymptotically distributed as χ^2 with q degrees of freedom. This theorem has several important applications.

(1) *Goodness-of-fit tests*

The full model is that each observation is fitted independently, usually by making the fitted value equal the observed value. For example, the log-likelihood for a sample from Poisson distributions with means $\mu_i(\theta)$ is

$$L = \sum_i (\mu_i - y_i \log \mu_i) + \text{constant terms},$$

and the difference between L_n, an exact fit with $\mu_i = y_i$, and L_p with fitted means $\hat{\mu}_i$, gives the criterion

$$\gamma = 2 \sum_i \left(\hat{\mu}_i - y_i - y_i \log\left(\frac{\hat{\mu}_i}{y_i}\right) \right),$$

which is asymptotically distributed as χ^2 with $(n - p)$ degrees of freedom.

The above test may be used only when the distribution is Poisson, binomial, or multinomial. With very small expectations, in particular with binary (0, 1) data, the test is not valid, and it is necessary to combine observations into groups. Karl Pearson recommended a minimum expectation of 5 in any class, but there is evidence that the test may be good enough if expectations exceed 1. The test cannot be used for the normal distribution unless the variance, σ^2, is known. In general, therefore, there is no goodness-of-fit test for regression models, although if an estimate of σ^2 is available analysis of variance may be used.

(2) *Analysis of deviance*

When a hierarchy of models is fitted, each model being a special case, omitting some parameters of the next model in the hierarchy, the differences in deviance may be set out as an analysis of deviance. If the numbers of parameters estimated in the sequence by $p_1, p_2, \ldots, p_k \, (= n)$, where the last model is the complete model with all expectations set equal to the observation, then the analysis may be set out as follows:

Source	Deviance change	Degrees of freedom
Model 2 versus Model 1	$2(L_{p_1} - L_{p_2})$	$p_2 - p_1$
Model 3 versus Model 2	$2(L_{p_2} - L_{p_3})$	$p_3 - p_2$
	...	...
Residual error	$2(L_{p_{k-1}} - L_n)$	$n - p_{k-1}$

For Poisson, binomial, and multinomial distributions each line may be tested against critical values of the appropriate χ^2 distribution. For normal errors the deviance criterion is simply the weighted sum of squared residuals (RSS), so that assuming a common unknown value of the variance σ^2 for each model, ratios of mean squares may be tested against critical values of the F distribution. The table then becomes an *analysis of variance*.

If the models to be compared cannot be expressed as a hierarchical sequence there is no satisfactory formal procedure for testing that one model is an improvement in the fit. However, it is common practice to compare the deviances for alternative models with the same number of parameters, such as curves of different mathematical form or different frequency distributions. It is not automatically sensible to accept the model that fits best, and it is advisable to examine the implications of such a choice in the context of the use to which the fitted model is to be put (see Cox, 1962).

An important class of applications occurs when the same model is fitted to several data sets, and some parameters may be assumed common to each set. The example of parallel curves has already been described in Section 1.1.2, and the general case has been called *parallel model analysis* (Ross, 1984).

(3) *Likelihood-based confidence regions*
The results of the analysis of deviance may be inverted by examining the set of parameter values $\hat{\theta}$ which satisfy a particular test of hypothesis. In particular, let τ be the MLEs for a model with p_2 parameters, and let τ^* be a value of θ in which p_1 parameters are estimated and the remaining $p_2 - p_1$ are assumed known. The critical values of $\hat{\theta}^*$ that satisfy the analysis of deviance test are such that $L(\hat{\theta}^*) - L(\theta) = \frac{1}{2}\chi^2$ on $(p_2 - p_1)$ degrees of freedom, or in the case of the analysis of variance,

$$\frac{\mathrm{RSS}(\theta^*) - \mathrm{RSS}(\hat{\theta})}{\mathrm{RSS}(\hat{\theta})} = \frac{n - p_2}{p_2 - p_1} F_c(p_2 - p_1, n - p_2).$$

These formulas define *confidence regions* for (θ) bounded by contours of $L(\theta)$ or $\mathrm{RSS}(\theta)$, surrounding the points representing the MLEs in p-dimensional parameter space.

These regions are discussed in detail in Section 3.2.1. They are known as *likelihood-based confidence regions* (Cox and Hinkley, 1974).

1.2.3. Alternative Methods of Estimation

Although maximum likelihood estimation is the most widely used method, if one includes all the applications of linear and nonlinear regression, there has been much theoretical discussion of alternative methods of estimation. These may be classified as:

(1) Simplified methods, providing unique estimates without iteration.
(2) Methods that employ additional information.
(3) Methods robust to gross errors in the data.

(1) *Simplified methods of estimation*
Before maximum likelihood estimation was developed a common type of estimation procedure was to find appropriate functions of the expectations, $E(y)$, which can be equated to corresponding functions of the observations; from p equations in p unknowns the parameters are estimated. For example, the *method of moments* (Pearson, 1894) equates theoretical and sample moments in order to fit frequency distributions. Thus the Neyman Type A distribution (see, e.g., Johnson and Kotz, 1969) has parameters m_1 and m_2, with mean $m_1 m_2$, and variance $m_1 m_2(1 + m_2)$ from the sample moments of which m_1 and m_2 may be estimated. Some curves can be fitted by selecting p individual ordinates (fitted values) or mean values, and solving the resulting nonlinear simultaneous equations. Others can be fitted by transforming the expectations to linear form, and fitting linear regression, ignoring the change in the error assumptions. Thus the rectangular hyperbola

$$E(y) = \frac{ax}{b + x}$$

is commonly fitted by writing

$$\frac{1}{y} = \frac{1}{a} + \frac{b}{ax}$$

and fitting a straight line for $1/y$ on $1/x$.

Such methods can be considered as equivalent to changing the weights attached to each observation, so that the solutions tend to coincide when the residuals are very small.

(2) *Use of additional information*
Bayes estimation uses Bayes' theorem to combine the likelihood with a prior distribution of the parameters, $p_1(\theta)$, to give a posterior probability,

$$p_2(\theta|y) = p_1(\theta) \times \text{likelihood}(\theta|y),$$

which may then be maximized. The posterior distribution of a particular parameter θ_1 is than obtained by integrating p_2 with respect to the other parameters. Sometimes the prior information may be considered as roughly equivalent to extra data modeled by the distribution $p_1(\theta)$, and the log-likelihood function is modified by an additive component which will alter the position of the optimum and tend to reduce the size of the confidence regions.

Ridge regression (Hoerl and Kennard, 1970) is normally applied only to linear multiple regression models, and is equivalent to placing extra hypothetical observations symmetrically about the centroid of the data, thus tending to reduce correlations between independent variables. The justification for this procedure is not clear, but there have been numerous papers on the subject.

(3) *Robust estimation*

When data are possibly subject to gross errors, methods have been proposed that give less weight to observations with the largest residuals. In the Princeton study (Andrews *et al.*, 1972) various weight functions were proposed, such as

$$w_i = 1 \qquad \text{if} \quad |y_i - \hat{y}_i| \le c,$$

$$w_i = \frac{c}{|y_i - \hat{y}_i|} \qquad \text{if} \quad |y_1 - \hat{y}_i| > c.$$

Such weights are functions of the parameters, but the method may still be considered a modification of maximum likelihood.

Similar robust results are obtained by minimizing the L_1-norm, $|\mathbf{y} - \hat{\mathbf{y}}(\boldsymbol{\theta})|$, where the method is equivalent to weighted least squares with weights $1/|y_i - \hat{y}_i|^{1/2}$, which also depend on parameters and reduce the influence of observations with large residuals.

The foregoing discussion suggests that findings for nonlinear estimation by maximum likelihood may also apply, with some modification, to other methods of estimation.

CHAPTER 2

Transformations of Parameters

2.1. What Are Parameters?

Mathematically, parameters are the unknown quantities θ in the likelihood function, $L(\theta)$, whose values $\hat{\theta}$ at the optimum are regarded as giving the best fit of the model to the data. There is an unlimited number of ways of writing the same model in terms of different sets of parameters. But the reasons for a particular parametrization of θ are not generally discussed in statistical texts.

Consider, for example, an experiment to determine the maximum point on a curve, $y = f(x)$, where the function $f(x)$ may be approximated by a quadratic polynomial. The simplest algebraic formulation of the model is the expression

$$f(x) = \beta_0 + \beta_1 x + \beta_2 x^2, \tag{2.1.1}$$

which is linear in the parameters β_0, β_1, and β_2 and which therefore may be fitted by multiple linear regression, assuming normally distributed errors. If the values of x are all positive the estimates of the parameters β are likely to be highly correlated, and subject to numerical inaccuracy. For accurate numerical work, orthogonal polynomials are recommended, and the function is rewritten

$$f(x) = \alpha_0 + \alpha_1(x - \bar{x}) + \alpha_2((x - \bar{x})^2 - \overline{(x - \bar{x})^2}), \tag{2.1.2}$$

so that parameters α are independently estimable. But the purpose of fitting the model is to estimate θ_1, the maximum value of $f(x)$, and θ_2, the corresponding value of x, which can appear directly as parameters if the model is written

$$f(x) = \theta_1 - \theta_3(x - \theta_2)^2, \tag{2.1.3}$$

where the remaining parameter θ_3 expresses the curvature at the maximum. The model is nonlinear in θ and therefore less suitable than (2.1.2) for estimation but form (2.1.3) is convenient if confidence regions for θ_1 and θ_2 are required.

This example illustrates three different types of parameter:

(1) *defining parameters* or quantities introduced as algebraic or polynomial coefficients in models, irrespective of the data;
(2) *computing parameters* or quantities adapted to improve the numerical properties of fitting the model to a particular data set; and
(3) *interpretable parameters* or quantities directly relevant to the particular application of the model.

There is no reason why a single set of parameters should be expected to fulfil all these roles simultaneously. Indeed, the number of interpretable quantities may differ from the number of parameters; we may be interested in only one such quantity (regarding the remaining parameters as nuisance parameters or specific parameters), or we may require estimates of several quantities, not necessarily independently estimable.

The relationships between the three parameterizations may be summarized in the following diagram:

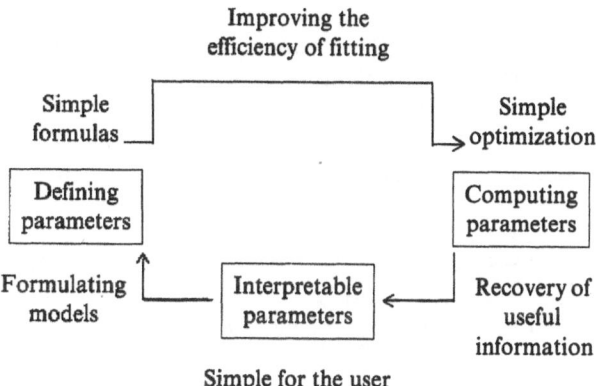

2.1.1. Transformations of Parameters

A transformation of parameters, $\phi = T(\theta)$, is a set of equations from which the p-component vector ϕ may be evaluated given the p-component vector θ. Its inverse, $\theta = T^{-1}(\phi)$, is not necessarily unique, or easily soluble.

The example in the previous section shows three equivalent parameter systems, β, α, and θ, connected by the following relationships:

$$
\left.
\begin{aligned}
\beta_0 &= \alpha_0 - x\alpha_1 + (\overline{x^2} - \overline{(x - \overline{x})^2})\alpha_2 = \theta_1^2 - \theta_3\theta_2^2, \\
\beta_1 &= \alpha_1 - 2\overline{x}\alpha_2 \qquad\qquad\qquad\quad = 2\theta_2 - \theta_3, \\
\beta_2 &= \alpha_2 \qquad\qquad\qquad\qquad\qquad\quad = -\theta_3.
\end{aligned}
\right\}
\qquad (2.1.4)
$$

Whichever system is used the model is the same, and the parameters estimated by maximum likelihood or least squares generate the same fitted curve with the same residual deviance. This identity of solutions assumes of course that the error distribution is unchanged; approximate methods of estimation (discussed in Section 1.2.3) that transform to a simpler form of model may get different results if the error distribution is also changed.

The essential difference between the parameter systems, β, α, and θ, is in the shape of the log-likelihood function in the neighborhood of the solution, especially its second- and higher-order derivatives, and the shape of its contours. These differences are illustrated by a two-parameter curve fitted to the data in Table 2.1: (1) with the defining parameters $E(y) = \theta_2 x/(x + \theta_1)$, and (2) with fitted values ϕ_1 (at $x = 3$) and ϕ_2 (at $x = 6$) which may be written as

$$E(y) = \frac{\phi_1 \phi_2 x}{(2\phi_1 - \phi_2) x + 6(\phi_2 - \phi_1)}.$$

Two parametrizations are compared in Fig. 2.1. The contours represent likelihood-based confidence regions corresponding to 60% and 95% critical values of the F distribution, but the important feature is their shape. In θ space the contours are elongated and asymmetrical, and the correlation between $\hat{\theta}_1$ and $\hat{\theta}_2$ is 0.959. In ϕ space the contours are nearly symmetrical ellipses, distorted slightly near the critical line $\phi_1 = \phi_2$, but much less elongated, giving a correlation between ϕ_1 and ϕ_2 of 0.296.

A second example is given in Fig. 2.2. The Neyman Type A distribution with parameters m_1 and m_2 is defined for nonnegative integers with probabilities

$$\left. \begin{array}{l} p(0) = \exp(-m_1(1 - e^{-m_2})), \\[2mm] p(r) = \dfrac{m_1 m_2}{r} e^{-m_2} \displaystyle\sum_{j=0}^{r-1} \frac{m_2}{j!} p(r - j - 1), \quad r \geq 1. \end{array} \right\} \qquad (2.1.5)$$

Table 2.1. Simulated data from rectangular hyperbola model, $E(y) = \theta_2 x(x + \theta_1)$, fitted by least squares, with $\hat{\theta}_1 = 2.549$, $\hat{\theta}_2 = 9.956$.

x	y	$E(y)$
1	2.7	2.81
2	4.4	4.38
3	5.3	5.38
4	6.6	6.08
5	6.5	6.59
6	6.3	6.99
7	7.7	7.30

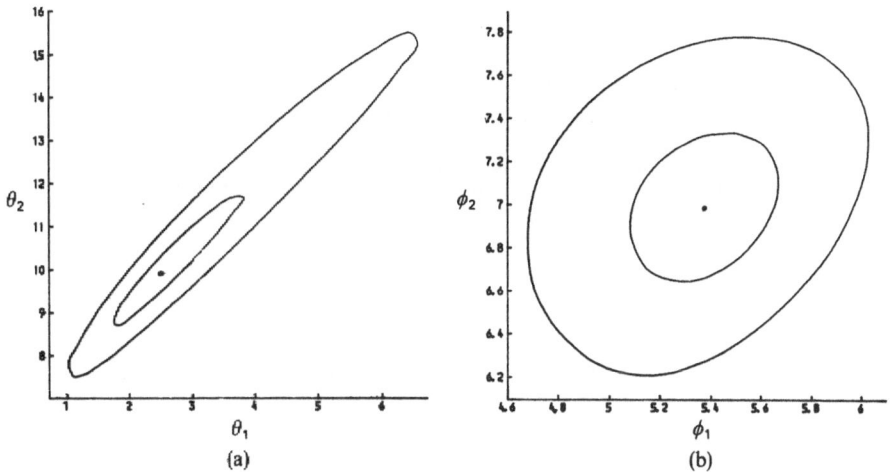

Fig. 2.1. Contours of residual sums of squares for the rectangular hyperbola fitted to the data of Table 2.1. Contour levels are 1.264 and 3.086 (see text). (a) Defining parameters, θ_1 and θ_2; (b) stable parameters, ϕ_1 and ϕ_2.

For the data given in Table 2.2, the figure gives likelihood contours: (1) in the space of m_1 and m_2, and (2) in the space of m and s, where $m = m_1 m_2$ is the expected mean of the distribution and $s^2 = m_1 m_2 (1 + m_2)$ is the expected variance. The critical contour in (m_1, m_2) space is curved, elongated, and is not even closed as m_1 approaches infinity. This is explained in (m, s) space by the fact that the contour crosses the critical curve $m = s^2$, below which points

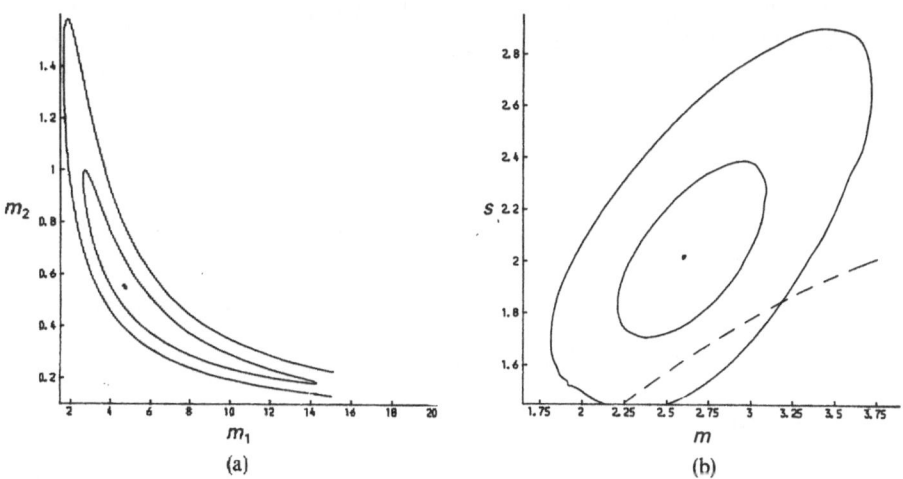

Fig. 2.2. Contours of residual log-likelihood for the Neyman Type A distribution fitted to the data of Table 2.2. Contour levels are 4.20 and 1.90. The outer contour is not closed. (a) Defining parameters, m_1 and m_2; (b) stable parameter s, $m = m_1 m_2$, and $s = \sqrt{m_1 m_2 (1 + m_2)}$. The broken line is the curve $s^2 = m$.

Table 2.2. Artificial data to illustrate likelihood
contours for the Neyman Type A distribution.
Maximum likelihood estimates are $\hat{m}_1 = 4.852$,
$\hat{m}_2 = 0.5394$, corresponding to $\hat{m} = 2.617$,
$\hat{s} = 2.007$.

No. in sample	Frequency	Expected frequency
0	3	4.0
1	8	6.1
2	7	6.3
3	4	5.1
4	2	3.6
5	3	2.3
6 or 7	2	2.0
8 or more	1	0.7

on the contour correspond to negative values of m_1 and m_2. These negative values are uninterpretable in terms of the Neyman Type A model (the distribution of individuals within groups, where the number of groups has the Poisson distribution with mean m_1, and the number of individuals obeys the Poisson distribution with mean m_2), but even so the computed probabilities in (2.1.5) are positive and valid for small values of r, sufficient to compute the likelihood for the data in Table 2.2.

2.1.2. Desirable Properties of Transformations

The two preceding examples illustrate, in two dimensions, nonlinear transformations which appear, visually, to transform contours into ellipses with centers close to the optimum. If a further linear transformation can then be found to render the parameters uncorrelated, then in a wide neighborhood of the optimum the parameters behave as if they were linear and orthogonal. The advantages of such transformations, if they exist, are:

(1) *Numerical*

(a) Rapid convergence of nonlinear optimization algorithms, as measured by the numbers of iterations required for convergence.
(b) Numerical accuracy of results because optima are well determined by finite difference methods.
(c) Successful convergence from a wide range of starting values.

(2) *Statistical*

(d) Independence of parameters.
(e) Unbiasedness of parameters when likelihood functions are symmetric about the solutions.
(f) Reliability of estimates of standard errors and confidence regions.

(3) *Interpretational*

(g) Identifiability of the limits of the region in which the defining parameters may be estimated, in terms of the solubility of the transformation equations relating the two sets of parameters.
(h) Ability to investigate differences and similarities between models fitted to the same data.
(i) Relationships between parameters and data statistics leading to an easy choice of initial estimates.

2.1.3. Stable Parameters

Parameters with advantages of the types listed in Section 2.1.2 are known as *stable parameters* (Ross, 1970). The term *stable* is not closely defined, nor is such a definition profitable, because it is not necessarily important to achieve all the objectives simultaneously. The term is used in a comparative sense, denoting a choice between different parametric representations of the same model. For example, ϕ *is more stable than* θ, if in the region of the optimum the log-likelihoood function is more closely approximated by a quadratic form (using the second-order terms in the Taylor series expansion about the optimum) in terms of ϕ rather than in terms of θ; the remainder term is a complicated function which may become large for values some distance from the optimum, but for the purposes of estimation and inference it is the neighborhood of the optimum that is important. Additionally, ϕ *is more stable than* θ if the estimates $\hat{\phi}$ are less intercorrelated than estimates $\hat{\theta}$, as measured, for example, by the conditioning of the dispersion matrices, or graphically by the orientation and eccentricity of the likelihood contours.

The term *stable* in its computing sense implies rapid convergence of iterative estimation procedures with no tendency to diverge. In its statistical sense, it implies a broad measure of agreement between different methods of estimation, whether approximate or based on optimization of some criterion such as the likelihood function. *Unstable* parameters are easily recognized. No parameter that can tend to infinity while the likelihood function tends to a finite value can be stable. In the model of Table 2.1 the parameters θ_1 and θ_2 cannot be stable, because the limiting form of the curve as θ_1 tends to infinity and $\theta_2 = \theta_1$ is the straight line

$$E(y) = \beta x,$$

which can be made to fit the data quite well. Clearly, data to which the full model is fitted can be very close to a straight line, or even be of opposite curvature so that the maximum likelihood estimates (MLEs) of θ_1 and θ_2 are both negative. The parameters θ are unstable both because of the asymmetry of the likelihood contours in Fig. 2.1 and because of their interdependence. No reasonable guess can be made about trial initial estimates of parameter θ_1 without some knowledge of parameter θ_2, whereas parameters ϕ_1 and ϕ_2 must clearly be close in some sense to their corresponding observed values.

The properties discussed in Section 2.1.2 are interrelated. If the discrepancy between the log-likelihood and its approximating quadratic is small, properties (a), (c), (e) and (f) hold. If the parameters are approximately independent (property (d)), then (b) and (i) hold. Property (g) depends on the finiteness of the range of likely values of each parameter and the ability to draw closed contours in two dimensions, for which property (f) is sufficient but not necessary. Property (h) is related to properties (d) and (i), because the ability to represent the model in terms of parameters which estimate independent aspects of the data allows the detailed differences between models to be more readily understood.

2.1.4. Parameter Transformations and the Modeling Process

Statistical modeling is an iterative process with three phases: (1) model selection, (2) model fitting, and (3) model evaluation. These may be illustrated by the example of the rectangular hyperbola (Fig. 2.1).

(1) *Selecting the model*
The suggestion that the model be written

$$E(y) = \frac{\theta_2 x}{x + \theta_1}$$

could be derived: (a) from looking at the scatter diagram and recognizing the underlying smooth curve as being possibly a hyperbola (although the exponential, $E(y) = \theta_2(1 - \exp(-\theta_1 x))$ might fit equally well)); or (b) from theory, such as the Michaelis–Menten equation for the relationship between the velocity of the chemical reaction (y) and the concentration (x) of a substrate, where θ_2 is the maximum velocity and θ_1 is the rate constant. The choice of model may also be dictated by tradition, or the recommendation of others.

(2) *Fitting the model*
Several methods have been proposed for fitting the model:

(a) Nonlinear regression with constant-variance errors using iterative optimization algorithms such as the Gauss–Newton method or its modifications, giving $\hat{\theta} = (2.549, 9.956)$.
(b) By linear regression in the form

$$E\left(\frac{1}{y}\right) = \frac{1}{\theta_2} + \left(\frac{\theta_1}{\theta_2}\right)\left(\frac{1}{x}\right),$$

with constant-variance errors for $1/y$. This gives different estimates, (2.519, 10.32), because the error assumption for ($1/y$) is not the same as that for y; in effect, small values of y are given unduly large weight compared with method (a).
(c) By making graphical estimates of the initial slope θ_2/θ_1 and the asymptote θ_2 using the plot of y against x. This method is used mainly to provide initial estimates for other methods.

(d) By separable estimation of the maximum θ_2 for each trial value of θ_1 (see Section 5.2.3). Given θ_1 the model becomes equivalent to a straight line through the origin on the transformed scale

$$z = \frac{x}{x + \theta_1}, \qquad E(y) = \theta_2 z,$$

for which the estimate of θ_2 is $\sum yz / \sum z^2$ and the residual sum of squares (RSS) is

$$R(\theta_1) = \sum y^2 - \frac{(\sum yz)^2}{\sum z^2}.$$

The function $R(\theta_1)$ is then minimized with respect to θ by an optimization algorithm.

(e) By transforming the parameters from θ to ϕ, as described in Section 2.1.1, to improve the performance of the optimization algorithm in (a), leading to $\hat{\phi} = (5.383, 6.988)$ which corresponds to $\hat{\theta} = (2.549, 9.956)$.

(f) By transforming the θ_1 parameter in (d) to improve the performance of the algorithm, for instance, by making $R(\theta_1)$ more symmetrical about its minimum. A suitable parameter is $\alpha = \phi_1/\phi_2$, which enables the model to be rewritten

$$E(y) = \phi_1 z \qquad \text{where} \quad z = \frac{x}{(2\alpha - 1)x + 6(1 - \alpha)}.$$

The function $R(\theta_1)$ and $R(\alpha)$ are compared in Fig. 2.3. The improvement is not perfect, but it is sufficient for the Gauss–Newton method to give three-figure accuracy in two iteration from any initial value of θ_1 in the range $(0.5, 1)$ compared with an equivalent performance of the same

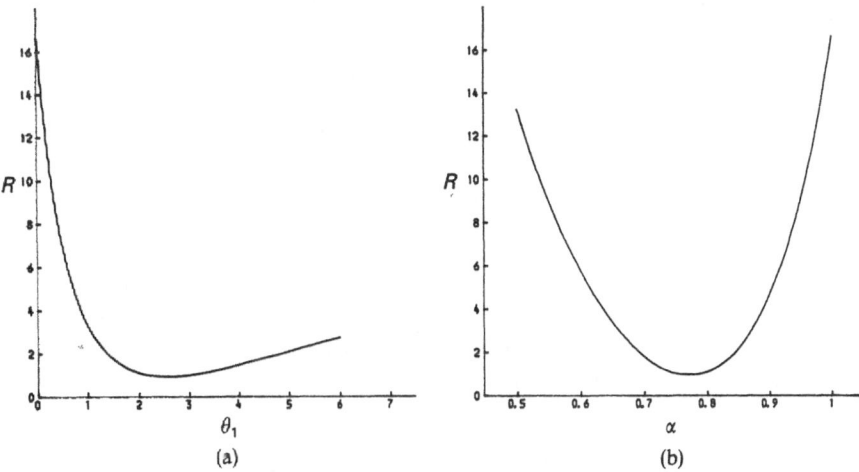

Fig. 2.3. Residual sum of squares as a function of the nonlinear parameter in the rectangular hyperbola fitted to the data of Table 2.1. (a) Defining parameter, θ_1; (b) stable parameter, $\alpha = \phi_2/\phi_1$.

algorithm in the scale of α only within the range $(0.7, 0.91)$ (corresponding to value of θ_1 in the range $(0.6, 4.6)$).

(3) *Interpreting the model*

Having fitted the model we may compute the predicted value of y for any value of x, the predicted value of x for any value of y up to θ_2, and the slope

$$\frac{dy}{dx} = \frac{\theta_1 \theta_2}{(x + \theta_1)^2}$$

for any value of x. We may also assign standard errors to these estimates using the approximate formula for the variance of a function of random variables (see Section 3.2). With greater effort exact confidence limits may be computed for any of these functions.

We may next question the assumptions about the model:

(a) Is it safe to assume that the errors are distributed normally and independently with equal variance, especially as the curve is constrained to pass through the origin? The effects of alternative weighting schemes or error distributions could be investigated.

(b) Is there evidence that a simpler model would be adequate, such as the straight line through the origin

$$E(y) = \beta x?$$

(c) Is there evidence that a more complex model could be fitted, either on theoretical grounds (such as the suggestion that more than one chemical reaction is taking place) or on empirical grounds (the observation that the data might fit better if there was a change of origin, or the asymptote was not horizontal)?

(d) Is there evidence that a different model with the same number of parameters might be preferred, such as the exponential model

$$E(y) = \theta_2(1 - \exp(\theta_1 x)),$$

or a quadratic polynomial through the origin? If the data set is one of a series it will be necessary to compare the estimates of θ_1 and θ_2 with those of other data sets. It is quite possible that some of these will encounter problems if the optimum values of ϕ lie too close to the critical lines $\phi_1 = \phi_2$ or $2\phi_1 = \phi_2$. In any case, it is clear from the contours in Fig. 2.1 that it will make little sense to treat θ_1 and θ_2 independently, or to use arithmetic means, whereas the ϕ parameters can be readily combined. A discussion of the analysis of groups of nonlinear models is given in Example 3.9 (Section 3.1.4).

2.1.5. Methods for Finding Stable Parameters

It is less easy to describe how to find stable parameters than to state their desirable properties. The examples given so far may seem contrived. The methods used may be appropriate in two dimensions where stability may be checked graphically, but how can they be generalized to more complex multi-parameter models? Is it possible to anticipate that a transformation will increase stability, or is it always a case of trial and error?

We can distinguish between *a priori* and *a posteriori* transformations.

A priori stable parameters (see Section 2.2) are those expected, on the basis of statistical reasoning or previous experience, to be sufficiently stable to allow the model to be fitted efficiently and accurately. They usually involve a function of the data.

A posteriori stable parameters (see Section 2.3) are those obtained with the full knowledge of the position of the optimum. Transformations within a parametric family can be optimized to reduce intercorrelations, or to eliminate cubic asymmetry in the log-likelihood in the neighborhood of the optimum. These refined parameters may then be used to estimate confidence regions and standard errors of functions.

2.2. *A Priori* Stable Parameters

Given data and a model expressed in terms of defining parameters θ, we seek parameters φ which are expected to resemble independent parameters in a linear regression. We seek parameters whose estimates will be approximately normally distributed with relatively small variance (relative to other possible parameters), and we require p such quantities approximately independent of each other.

The most fruitful general principle is as follows: examine the *descriptive statistics of the data set* and use as parameters the corresponding descriptive statistics of the expectations. If these quantities are algebraically too complicated to use, replace them by simpler expressions which have a similar effect. How this works in practice depends on context, but here are some examples.

2.2.1. Frequency Distributions: Moments and Percentiles

For discrete distributions such as the Poisson and its generalizations (negative binomial, Neyman Type A, Poisson–Pascal, etc.) the expected low-order moments are simple descriptive statistics that are functions of the parameters and individually suitable as stable parameters. But when the correlation between mean and variance is large it may be better to use, instead of the variance, a combined function such as

$$\frac{1}{k} = \frac{\text{variance} - \text{mean}}{(\text{mean})^2},$$

which for the negative binomial distribution is always independent of the mean. When most of the observations are zero there is further information in the zero frequency, p_0, which may be used as a parameter, as pointed out by Anscombe (1950). For continuous distributions similar considerations apply. For symmetric normal-like samples the expected mean and variance (or standard deviation, whose distribution is less skew) are very stable, but other statistics such as percentiles may be more useful than the higher moments which are poorly estimated in small samples.

The distinction between the proposal to use expected moments as parameters and the use of sample moments themselves (as in Pearson's method of moments) should be clear. The method of moments is an approximate method that equates particular functions of parameters to their sample equivalents, without regard to the suitability of those functions in terms of the amount of independent information available, leading to some cases to substantial inefficiency compared with MLEs.

2.2.2. Curve Fitting: Orthogonal Polynomial Analogies

The series of orthogonal polynomials illustrates one possible system of obtaining stable parameters. The parameters depend on the distribution of the independent variable, x, in the expression

$$E(y) = f(x, \theta) = \sum_{k=0}^{N}{}' \alpha_k P_k(x),$$

where $P_k(x)$ is a polynomial of degree k such that $\sum_i P_j(x_i)P_k(x_i) = 0$ for all $j = k$. The parameters α_k represent, in sequence,

  α_0 the mean height of the curve,

  α_1 the overall slope,

  α_2 the curvature compared with a straight line,

  α_3 a measure of asymmetry compared with the fitted quadratic.

Each of these parameters measures a contrasting feature of the data set, and orthogonality is achieved by an algorithm that depends only on x, not on y or the fitted parameters.

To analyze nonlinear models in this way it would be necessary to abandon any attempt to maintain strict orthogonality while following the same outline. For example, the exponential curve

$$E(y) = \alpha + \beta \exp(-kx) \tag{2.2.1}$$

tends to a straight line as $k \to 0$, so for data not very different from a straight line it would be natural to look for a single parameter to express the non-linearity; not an estimate of the parameter k in (2.2.1) which is highly correlated with estimates of α and β, but k in

$$E(y) = \alpha' + \beta'(f(k) - \overline{f(k)}), \tag{2.2.2}$$

where

$$f(k) = \frac{1 - \exp(-k(x - x_m))}{kx_m},$$

and is suitably chosen working origin of x. If k is close to zero, x_m tend to $\bar{x}$, but for larger k it is better to use a value towards the end of the range where the curve is steepest (i.e., away from the asymptote). Then equation (2.2.2) expresses through α' the mean height of the curve, and through β' the approximate mean slope; specifically, β' is the slope at $x = x_m$, as illustrated in Fig. 2.4.

Similar formulations are possible for the general rectangular hyperbola by writing

$$f(\delta) = \frac{x - x_m}{1 + \delta(x - x_m)} \tag{2.2.3}$$

and the straight line + exponential,

$$E(y) = \alpha + \beta \exp(-kx) + \gamma x, \tag{2.2.4}$$

which can be rewritten

$$E(y) = \alpha' + \beta'(x - \bar{x}) + \gamma' g(k), \tag{2.2.5}$$

where

$$g(k) = f(k) - \beta_1(k)(x - \bar{x}) - \beta_0(k),$$

and

$$f(k) = \frac{1 - k(x - x_m) - \exp(-k(x - x_m))}{k^2},$$

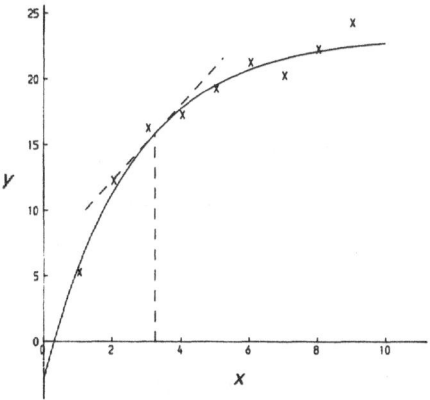

Fig. 2.4. Stable parameters for a fitted exponential curve. The data of Table 2.3, together with the fitted curve and the tangent at $x = 3.244$, are illustrated. Parameters are: (a) the mean value of y, (b) the slope of the tangent, and (c) the nonlinearity of the curve, represented by the exponential coefficient, k.

where γ' is the second derivative at $x - x_m$ and k represents the departure from the quadratic polynomial model. The functions $\beta_1(k)$ and $\beta_0(k)$ are linear regression coefficient of $f(k)$ on $(x - \bar{x})$.

There is clearly a limit to the convenience of such formulations, but when faced with a multiparameter curve which is difficult to fit, it will always be worthwhile investigating transformations that use low-order derivatives at some intermediate point in the x range.

Choice of x_m *a priori* is only approximate, and if need be can be adjusted *a posteriori* to improve stability.

2.2.3. Stable Ordinates: Curves Through Given Points

A completely different approach is to consider the general p-parameter curve as determined by p suitably spaced points on the curve, the *ordinates* (y values) at selected values of $x = \alpha_1, \ldots, \alpha_p$. Thus, if the model is

$$E(y) = f(x, \theta), \tag{2.2.6}$$

then the transformation from θ to ϕ is determined by the simultaneous equations

$$\phi = f(\alpha, \theta). \tag{2.2.7}$$

Choice of α is dictated partly by the range of data values, but also by the need to be able to solve equations (2.2.7) for θ in terms of ϕ. This approach has already been used in Fig. 2.1.

For polynomials of order $p - 1$ this is straightforward, and the model becomes

$$E(y) = \sum_{i=1}^{p} \phi_i \prod_{j=i} \left(\frac{x - \alpha_j}{\alpha_i - \alpha_j} \right). \tag{2.2.8}$$

For example, the quadratic becomes

$$E(y) = \phi_1 \frac{(x - \alpha_2)(x - \alpha_3)}{(\alpha_1 - \alpha_2)(\alpha_1 - \alpha_3)} + \phi_2 \frac{(x - \alpha_1)(x - \alpha_3)}{(\alpha_2 - \alpha_1)(\alpha_2 - \alpha_3)}$$
$$+ \phi_3 \frac{(x - \alpha_1)(x - \alpha_2)}{(\alpha_3 - \alpha_1)(\alpha_2 - \alpha_2)}. \tag{2.29}$$

The values of α_j, which make the ϕ parameters orthogonal, turn out to be the roots of any polynomial of order p orthogonal to the polynomials of lower order with respect to the data. For, if the polynomial

$$P_p(x) = \prod_{j=i}^{p} (x - \alpha_j) \tag{2.2.10}$$

is orthogonal to all lower-order polynomials, then the sum of the products of variables in the regression equations (2.2.8) will be of the form

$$\sum P_p(x)\, Q_{p-2}(x), \qquad (2.2.11)$$

where $Q_{p-2}(x)$ is a polynomial of order $p - 2$, and the products vanish.

For nonlinear curves such as the exponential curve (2.2.1) it is possible to find *a posteriori* values of α for which the ϕ parameters are almost uncorrelated. More useful, *a priori* are values of α which make equations (2.2.7) easily soluble. This conditon conflicts to some extent with the condition of independence, but by a suitable choice with respect to the observed x values, reasonably widely spaced values spanning the range give satisfactory results in most cases. Here are some examples:

(i) *Exponential curve*

$$E(y) = \alpha + \beta \exp(-kx). \qquad (2.2.12)$$

Choose three equally spaced values of x. Then

$$\frac{\phi_3 - \phi_2}{\phi_2 - \phi_1} = \exp(-k(x_3 - x_2)), \qquad (2.2.13)$$

whence k and the remaining parameters may be evaluated. In particular,

$$\alpha = \phi_3 - \frac{(\phi_3 - \phi_2)^2}{\phi_1 + \phi_3 - 2\phi_2}, \qquad (2.2.14)$$

an expression familiar in numerical analysis as Aitken's acceleration formula.

(ii) *Logistic curve*

$$E(y) = \frac{\gamma}{1 + \exp(-\beta(x - \mu))}. \qquad (2.2.15)$$

This curve is the reciprocal of the exponential, hence the parameters can be found by replacing ϕ_j by $1/\phi_j$ in expressions (2.2.13) and (2.2.14).

(iii) *Gamma curve*

$$E(y) = \alpha x^\beta \exp(-\gamma x), \qquad x > 0. \qquad (2.2.16)$$

Choose

$$x_2 = 2x_1, \qquad x_3 = 4x_1.$$

Then

$$\gamma = \frac{1}{x_1} \log \frac{\phi_2^2}{\phi_1 \phi_3}, \qquad \beta = \log \frac{\phi_2^3}{\phi_3 \phi_1^2} \bigg/ \log 2,$$

whence α follows.

(iv) *Sum of exponentials*

$$E(y) = \beta \rho^x + \gamma \sigma^x. \qquad (2.2.17)$$

Choose equally spaced points $\alpha = 1, 2, 3, 4$ after scaling x. Then

$$\rho + \sigma = \frac{\phi_1\phi_4 - \phi_2\phi_3}{\phi_1\phi_3 - \phi_2^2},$$

$$\rho\sigma = \frac{\phi_2\phi_4 - \phi_3^2}{\phi_1\phi_3 - \phi_2^2}, \tag{2.2.18}$$

whence ρ and σ may be solved as (real, positive) roots of the equation

$$t^2 - (\rho + \sigma)t + \rho\sigma = 0,$$

and β and γ then follow.

Alternatively, we may use $\alpha = 1, 2, 3, 5$, in which case

$$(\rho + \sigma)^2 = \frac{\phi_1\phi_4 - \phi_3^2}{\phi_1\phi_3 - \phi_2^2},$$

$$\rho\sigma = \frac{\phi_2(\rho + \sigma) - \phi_3}{\phi_1}, \tag{2.2.19}$$

which yields a similar quadratic for ρ and σ, provided the right-hand side of (2.2.19) is positive. For the general form

$$E(y) = \alpha + \beta\rho^x + \gamma\sigma^x, \tag{2.2.20}$$

five equally spaced points are required, and after taking first differences

$$\phi_2 - \phi_1 = \beta(\rho - 1) + \gamma(\sigma - 1),$$

$$\phi_3 - \phi_2 = \beta(\rho - 1)\rho + \gamma(\sigma - 1)\sigma, \quad \text{etc.}$$

the equations may be solved as in (2.2.18).

Three exponentials

$$E(y) = \beta\rho^x + \gamma\sigma^x + \delta\tau^x \tag{2.2.21}$$

require six equally spaced points, and if we write

$$S_0 = \phi_3(\phi_3^2 - \phi_2\phi_4) - \phi_4(\phi_2\phi_3 - \phi_1\phi_4) + \phi_5(\phi_2^2 - \phi_1\phi_3),$$

$$S_1 = \phi_4(\phi_3^2 - \phi_2\phi_4) - \phi_5(\phi_2\phi_3 - \phi_1\phi_4) + \phi_6(\phi_2^2 - \phi_1\phi_3),$$

$$S_2 = \phi_4(\phi_3\phi_4 - \phi_2\phi_5) - \phi_5(\phi_3^2 - \phi_1\phi_5) + \phi_6(\phi_2\phi_3 - \phi_1\phi_4),$$

$$S_3 = \phi_4(\phi_4^2 - \phi_3\phi_3) - \phi_5(\phi_3\phi_4 - \phi_2\phi_5) + \phi_6(\phi_3^2 - \phi_2\phi_4),$$

then ρ, σ, and τ are roots of the cubic

$$S_0 t^3 - S_1 t^2 + S_2 t - S_3 = 0,$$

all of which must be real and positive to be interpretable.

(v) *Power law functions*
The generalized logistic

$$E(y) = \frac{\gamma}{(1 + \tau \exp(-\beta(x - \mu)))^{1/\tau}} \qquad (2.2.22)$$

may be solved, given τ as the ordinary logistic through three points ϕ_1^{τ}, ϕ_2^{τ}, and ϕ_3^{τ}, as in section (ii) above. Direct solution of the model through four points is complicated and needs iterative techniques.

The generalized hyperbola

$$E(y) = \alpha + \frac{\beta}{(1 + \gamma x)^{\delta}} \qquad (2.2.23)$$

is more awkward, as the parameters γ and δ are very unstable, so it may be preferable to use trial values of α and solve the equations

$$\log(\phi_j - \alpha) = \log \beta - \delta \log(1 + \gamma \alpha_j)$$

numerically for γ.

These examples show that procedures based on stable ordinates are not necessarily straightforward numerically. However, that does not detract from their stability.

2.2.4. Complex Multiparameter Models

Frequency distributions and nonlinear curves are relatively simple models in terms of independent variables. Multiway tables, multivariate regression (surfaces and higher-dimensional relationships), and multivariate frequency distributions are more complex and require many more parameters. How then can transformations be found without leading to intractable algebra? The same principles apply as before; if the model is not overparametrized with regard to the data there should be a sufficient number of independent contrasts to represent a stable parameter system. Even if a complete set cannot be found, every stable parameter represents an improvement on an unstable defining parameter system. Here are some examples.

(i) *Multivariate frequency distributions*
The low-order sample moments of a multivariate frequency distribution may be simple functions of the parameters. For example, to fit a mixture of two bivariate normal distributions with different means but common dispersion matrix,

$$f(x_1, x_2) = \alpha\Phi(\mu_{11}, \mu_{12}, \sigma_1, \sigma_2, \rho) + (1 - \alpha)\Phi(\mu_{21}, \mu_{22}, \sigma_1, \sigma_2, \rho), \quad (2.2.24)$$

a model with eight parameters, we note that

$$E(x_1) = \alpha\mu_{12} + (1 - \alpha)\mu_{21},$$

$$E(x_2) = \alpha\mu_{12} + (1 - \alpha)\mu_{22},$$

$$\text{Var}(x_1) = \alpha(1 - \alpha)(\mu_{11} - \mu_{21})^2 + \sigma_1^2,$$
$$\text{Var}(x_2) = \alpha(1 - \alpha)(\mu_{12} - \mu_{22})^2 + \rho\sigma_1\sigma_2,$$
$$\text{Cov}(x_1, x_2) = \alpha(1 - \alpha)(\mu_{11} - \mu_{21})(\mu_{12} - \mu_{22}) + \rho\sigma_1\sigma_2,$$

and if these five expressions are treated as parameters then the equations may be solved for the means μ_{ij} and ρ in terms of α, σ_1, and σ_2. The latter parameters are constrained to lie in the ranges $0 < \alpha < 1, 0 < \sigma_1^2 < \text{Var}(x_1), 0 < \sigma_2^2 < \text{Var}(x_2)$, which gives warning of data for which the model is unsuitable.

(ii) *Groups of curves or other models*
When fitting the same form of curve to different data sets (see Section 1.1.2) it is useful to fit models in which some parameters are common to each set and other parameters specific to each set. The model becomes

$$E(y_{ij}) = f(x_i, \theta, \phi_j),$$

where the parameters θ are common and ϕ_j are specific. For example, to fit exponential curves of the same shape but different origin in the x axis, the model can be written

$$E(y_j) = \alpha(1 - \exp(-k(x - \gamma_j))), \tag{2.2.25}$$

for which stable parameters could be three ordinates in the first set of data and one ordinate for each subsequent set. This accords with the use of the model for calibration; estimating the displacement in (x) necessary to bring the data sets to fit a common curve.

Alternatively, the parameter k alone may be common to all curves, and the model becomes

$$E(y_j) = \alpha_j + \beta_j \exp(-kx), \tag{2.2.26}$$

which can be represented in terms of three ordinates of the first set and two for each subsequent set.

(iii) *Surfaces and multivariate regression*
Stable parameters for surface fitting depend crucially on the design of the data with respect to the independent variables x_1 and x_2. For points equally spaced on a rectangular grid it may be convenient to work with mean expected values along lines parallel to either axis, rather than single ordinates at particular positions. Sometimes the model can be regarded as the sum or product of curves, such as

$$E(y) = \alpha + \beta_1 \exp(-k_1 x_1) + \beta_2 \exp(-k_2 x_2), \tag{2.2.27}$$

or

$$E(y) = \alpha(1 - \beta_1 \exp(-k_1 x_1))(1 - \beta_2 \exp(-k_2 x_2), \tag{2.2.28}$$

which, for given x_2, is exponential in x_1 and vice versa.

(iv) *Multiway tables*

Some multiway tables are essentially samples from discrete frequency distributions, but cell expectations may be modified by row and column effects. For example, the life distributions of birds are sometimes estimated from data relating to the ringing of young birds and the return of the rings in subsequent years (see, e.g., Brownie and Robson, 1976). Data consists of n_{ij} rings from birds of age i returned in year j and N_k, the total number of rings issued in year $k = j - i$.

If π_i is the probability that a bird alive at age i is still alive at age $i + 1$, then the expected number dying at age r is proportional to $\pi_0 \pi_1 \ldots \pi_{r-1}(1 - \pi_r)$. However, only a small proportion, γ, of rings are returned, and for some species there are marked effects of calender year, either through the changed rates of mortality of birds of all ages, or changes in the rate of return, γ. Estimation of parameters in these models is considerably eased by transforming to stable parameters which roughly estimate the marginal totals in the table, rather than the ratios π_i which are correlated with γ and with each other. It is not necessary to attempt to extract stable parameters from the complicated algebraic expressions involved; this may not be possible. It is sufficient to note that the parameters

$$\phi_1 = \gamma(1 - \pi_0),$$
$$\phi_2 = \gamma\pi_0(1 - \pi_1),$$
$$\phi_3 = \gamma\pi_0\pi_1(1 - \pi_2),$$

are likely to be stable even for complex models in which the effects of year of return and year of issue are included.

Other examples of multiway contingency tables occur in genetical research, and in the study of epidemics and related phenomena where data are classified by state and time. The algebra of stability is not usually simple, but where alternative parametizations are possible it should be easy to recognize which ones are likely to be more stable.

2.3. *A Posteriori* Stable Parameters

If no simple transformation to stable parameters is available, a direct fit must be attempted. By examining the likelihood function in the neighborhood of the solution it is possible to find stabilizing transformations *for a particular data set*. These may suggest suitable transformations for other data sets, especially when they can be interpreted in terms of estimable statistics of the data. By examining several different data sets in this way the pattern may become clear.

A priori stable parameters (Section 2.2) usually involve *working constants* such as x_m in formula (2.2.2), or the choice of values of α for stable ordinates (formula (2.2.7)). To improve stability *a posteriori* it is possible to adjust these working constants using the information in the fitted parameters and dispersion matrix. Additionally, the parameters may be transformed, individually or in combination, to improve the symmetry of the likelihood function.

The problem may be approached:

(a) *analytically*, by attempting to reduce the off-diagonal terms in the information matrix to zero, and further to reduce higher-order terms in the Taylor expansion of the log-likelihood;
(b) *graphically*, by plotting contours of the log-likelihood function, and the remainder function after subtracting the quadratic approximation; and
(c) *numerically*, by observing the paths of convergence of optimization algorithms.

2.3.1. Analytical Methods for *A Posteriori* Stability

Given the log-likelihood function in terms of $(\theta - \hat{\theta})$

$$L(\theta) = L(\hat{\theta}) + \tfrac{1}{2}(\theta - \hat{\theta})'V^{-1}(\theta - \hat{\theta}) + R(\theta - \hat{\theta}), \qquad (2.3.1)$$

where $R(\theta - \hat{\theta})$ consists of cubic and higher-order terms, then the effect of a transformation from θ to ϕ in terms of the Jacobian matrix,

$$J = \left(\frac{\partial \phi_j}{\partial \theta_j}\right)_{\phi = \hat{\phi}},$$

is to rewrite the log-likelihood as

$$L(\phi) = L(\hat{\phi}) + \tfrac{1}{2}(\phi - \hat{\phi})'(J^{-1})'V^{-1}J^{-1}(\phi - \hat{\phi}) + R^*(\phi - \hat{\phi})$$
$$= L(\hat{\phi}) + \tfrac{1}{2}(\phi - \hat{\phi})'V_\phi^{-1}(\phi - \hat{\phi}) + R^*(\phi - \hat{\phi}),$$

where $V_\phi = JVJ'$ is the asymptotic dispersion matrix of the ϕ parameters and R^* is the remainder function.

(i) *Uncorrelated parameters*
For multiparamater models it is not usually feasible to obtain either analytical equations for the values of the working constants that produce a diagonal matrix V_ϕ, or useful expressions for reducing R^*. Since the number of terms to be reduced increases more rapidly than the number of constants to adjust, a perfect solution may not exist. That the back transformations, $\phi \to \theta$, be easily computed reduces the possibilities still further. For example, the transformation to stable ordinates for the double exponential curve (2.2.20) requires five ordinates with their ten correlations close to zero, yet to allow the back transformations we require equal spacing, hence only two quantities may be varied, the spacing and the position of the first ordinate.

Thus, given the fictitious data in Table 2.3, the least correlated ordinates are at $x = 1.141, 3.429, 7.468$, as found by minimizing the sum of the squared

Table 2.3. Data to which the model, $E(y) = \alpha + \beta \exp(-kx)$, is fitted, $\alpha = 23.288$, $\beta = -26.138$, $k = 0.3871$.

x	1	2	3	4	5	6	7	8	9
y	5	12	16	17	19	21	20	22	24

covariances. These values are not equally spaced; the best solution for equally spaced ordinates, at approximately $x = 1.36, 4.76, 8.16$, has $r_{23} = 0.114$.

(ii) *Reducing asymmetry*
One-parameter studies (Wedderburn, 1972) of transformations to reduce like-lihood asymmetry have used the condition

$$\frac{\partial^3 \theta}{\partial \phi^3} = 0$$

to give, for example, the appropriate power of θ in a family of power transfor-mations.

For a single sample from a normal distribution, the parametrization, $\phi = (\sigma^2)^{1/\alpha}$, for unknown α has log-likelihood

$$L(\phi) = \frac{\sum (x - \mu)^2}{2\phi^\alpha} \times \frac{n\alpha}{2} \log \phi.$$

The gives

$$2\frac{\partial^3 L(\phi)}{\partial \phi^3} = -\sum (x - \mu)^2 (\alpha(\alpha + 1)(\alpha + 2) \phi^{\alpha-3}) + 2n\alpha\phi^{-3} = 0,$$

and since $\hat{\phi}^\alpha = \sum (x - \mu)^2/n$ we obtain the equation for α

$$(\alpha + 1)(\alpha + 2) = 2,$$

whence $\alpha = -3$. The inverse cube root transformation is not, however, suffi-cient to produce elliptical contours, as is shown in Fig. 2.5. Similar methods show that for the Poisson distribution with mean μ, $\phi = \mu^{1/3}$, and for the binomial distribution with parameter p,

$$\phi = I(p, \tfrac{1}{3}, \tfrac{1}{3}) = \int_0^P \frac{(x(1 - x))^{2/3} \, dx}{B(\tfrac{1}{3}, \tfrac{1}{3})},$$

an incomplete beta function.

The importance of these results is that they may be combined with the stable ordinate methods to improve symmetry in the likelihood when the errors in the model are not normal. For example, if for the hyperbola in Fig. 2.1 we assumed the Poisson distribution (suitably scaled), the cube roots of the stable ordinates would be likely to produce more symmetric contours than would the ordinates themselves.

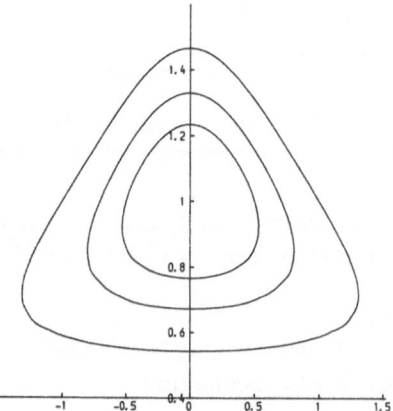

Fig. 2.5. Likelihood contours for normal samples. Contour levels are $0.25n$, $0.5n$, and n where n is the sample size. Parameters are $(\mu - \bar{x})/s$ and $(\sigma/s)^{-2/3}$.

2.3.2. Graphical Methods for *A Posteriori* Stability

Graphical methods (see also Chapter 4) are restricted to operations on one or two parameters at a time, yet can be very useful in the multiparameter case. The graphs used are:

(i) curves of log-likelihood plotted against θ_i, or any straight-line cross sections of parameter space defined by two end-points;
(ii) contours of equal log-likelihood in any plane cross section of parameter space, defined by three corners of a parallelogram; or
(iii) a surface-viewing plot made up of intersecting systems of curves, observed from a suitable viewpoint, with hidden segments suitably omitted or differently denoted.

For demonstration purposes solid surfaces may be constructed in plaster or similar materials, or from pieces of card cut to the shape of the one-dimensional cross sections and slotted together, as in Fig. 2.6.

The first two plots are the most useful in detecting asymmetry (and other more serious deviations from quadratic from, such as multiple minima, plateaux, or curved valleys). If the deviations are only slight they may be highlighted by plotting the remainder function in equation (2.3.1),

$$R(\theta - \hat{\theta}) = L(\theta) - \tfrac{1}{2}(\theta - \hat{\theta})' \, V^{-1}(\theta - \hat{\theta}), \qquad (2.3.2)$$

an example of which is shown in Fig. 2.7. These functions are of two types according to the number of real roots of the leading cubic expression. If there is only one real root, the function R is divided into one positive and one negative zone, the positive zone corresponding to likelihood contours closer together than expected, and the negative zone to contours further apart. If there are three real roots there are six zones, three positive and three negative, indicating that the likelihood contours are curved as well as asymmetric.

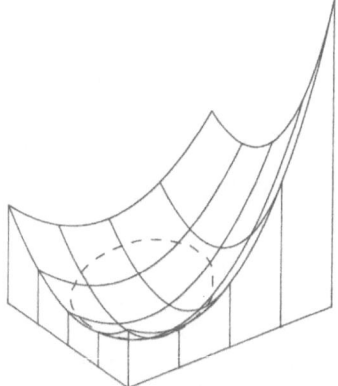

Fig. 2.6. Surface view of likelihood function showing likelihood contours as broken lines.

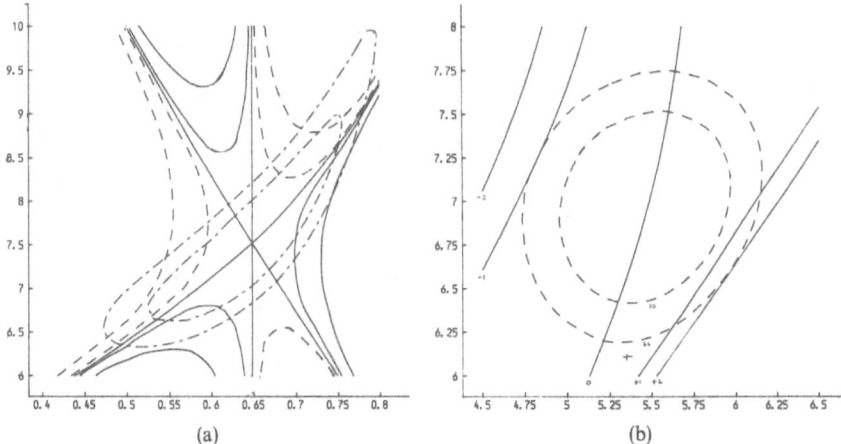

Fig. 2.7. Contours of remainder functions and likelihood functions, showing deviations from elliptical form. (a) Exponential model, $E(y) = \theta_2(1 - \theta_1^x)$, fitted to the data of Table 2.1; (b) the same, transformed to stable parameters representing values of $E(y)$ at $x = 2.5$ and $x = 5$.

In studying likelihood functions for three or more parameters it is necessary to look at two parameters at a time, e.g., cross sections including the solution. The pair with the least stable representation should be transformed first.

2.3.3. Numerical Methods for *A Posteriori* Stability

Numerical information about the log-likelihood function is available in several ways:

(i) during optimization, by examining the path of convergence in parameter space for linear or nonlinear relationships;

(ii) from the dispersion matrix at the solution;

(iii) by finding points on a particular likelihood contour, and measuring asymmetric and nonlinearity from the positions of extrema.

These methods are amplified below.

(i) *Convergence sequence during optimization*

If convergence is slow, the sequence of parameter values that corresponds to a decreasing series of log-likelihoods at the end of each iteration of a quadratic-ally convergent optimization algorithm should be examined carefully. Fast convergence is one of the surest criteria of a stable parameter system, while slow convergence indicates the presence of nonlinear relationships between parameters described geometrically (in two dimensions) as "curved descending valleys." For example, the three-parameter model

$$E(y) = \alpha + \beta \exp(-kx),$$

fitted to the data $x = 1, \ldots, 7$, $y = 6, 11, 13, 14, 16, 17, 19$, shows in Table 2.4 the pattern of convergence using the Newton–Raphson algorithm.

If we can find a relationship between α, β, and k that is almost invariant for each data set, it will point to a possible stable parameter. From the previous discussion (Section 2.3.1) it is not surprising that the function

$$\phi_3 = \alpha + \beta \exp(-3k),$$

(which takes the values 13.18, 12.77, 12.87, 12.71, 12.74, 12.71), and

$$\phi_4 = \alpha + \beta \exp(-4k),$$

(which takes the values 15.00, 14.69, 14.85, 14.72, 14.76, 14.73) are less variable than the original parameters.

(ii) *Analysis of the dispersion matrix*

Linear transformations will not necessarily help, as they do not theoretically affect the nonquadratic nature of the log-likelihood function. However, reduction in ill-conditioning of the information matrix can have an important effect on convergence. The most stable linear transformation is obtained by principle component analysis of the dispersion matrix. The vector corresponding to the smallest eigenvalue is the linear function of parameters with the smallest variance. For example, in the example above we find that the disperson matrix of (α, β, k) is

$$\begin{pmatrix} 7.1625 & -2.9862 & -0.27292 \\ -2.9862 & 2.8469 & 0.08729 \\ -0.27292 & 0.08729 & 0.0011025 \end{pmatrix},$$

Table 2.4. Parameter values from successive iterations
indicating a nonlinear relationship between the
parameter estimates.

Interation	α	β	k	RSS
2	18.31	−19.19	0.4397	3.988
3	19.45	−18.48	0.3393	3.088
4	20.05	−18.93	0.3231	2.658
5	20.73	−19.03	0.2880	2.569
6	20.87	−19.14	0.2853	2.546
9 (Final)	21.06	−19.20	0.2775	2.543

whose smallest eigenvalue is 0.000187 with the corresponding eigenvector proportional to (1, 0.368, 22.22) which happens to correspond fairly closely with the first derivatives of ϕ_3 with respect to the parameters (1, 0.435, 25.05).

It is a simple matter to obtain orthogonal linear functions of the parameters, for example, from the Choleski decomposition into triangular form. For, if A is a $p \times p$ matrix, then the dispersion matrix of $A\theta$ is $A'VA$, and if $A'VA$ is diagonal, then $A\theta$ is an orthogonal transformation of θ. The Choleski decomposition computes an upper triangular matrix T where $TT' = V^{-1}$, hence $T\theta$ has variance $T'VT = T'(TT')^{-1}T = I$. In the above example

$$T = \begin{pmatrix} 3.325 & 1.273 & 72.23 \\ 0 & 0.681 & -5.39 \\ 0 & 0 & 9.52 \end{pmatrix},$$

in which the first row is incidentally proportional to (1, 0.383, 21.72) corresponding approximately to ϕ_3, and the second row corresponds to the slope at $x = 2.1$.

Orthogonalization methods are not unique, and their main role is to simplify the computations rather than to suggest interpretations for the linear combinations of parameters so found.

(iii) *Extreme points on likelihood contours*
Methods for finding numerical values for extreme points on a likelihood contour are described in Section 5.2.4. Using these methods the effect of parameter transformations on the symmetry of the likelihood-based confidence intervals can be assessed. A simple scale-free measure of symmetry is the ratio

$$\frac{(\theta_{max} + \theta_{min})/2 - \hat{\theta}}{\theta_{max} - \theta_{min}},$$

where θ_{max} and θ_{min} are limit values of a particular parameter. A fixed contour must be used in making the comparison, as the ratio is not independent of the contour chosen.

2.4. Theoretical Justification for Stable Parameters Using Deviance Residuals

Deviance residuals (McCullagh and Nelder, 1982; Ross, 1982) are defined in terms of the contribution of each observation to the residual deviance. For a single observation y_i, with fitted expectation μ_i and distribution $p(y_i|\mu_i)$, the deviance residual is defined as

$$e_i = \text{sign}(y_i - \mu_i)\{2(\log p(y_i|y_i) - \log p(y_i|\mu_i))\}^{1/2}.$$

Clearly, $e_i = 0$ when $\mu_i = y_i$, and the relative deviance when observations are independent is $\sum e_i^2$. The relative log-likelihood is thus expressible as a sum of squares.

Some examples of deviance residuals are as follows:

(a) Normal distribution,

$$e_i = \frac{y_i - \mu_i}{\sigma_i}.$$

(b) Poisson distribution,

$$\tfrac{1}{2} e_i^2 = y_i \log\left(\frac{y_i}{\mu_i}\right) - y_i + \mu_i.$$

(c) Binomial y_i/n_i,

$$\tfrac{1}{2} e_i^2 = y_i \log_i\left(\frac{y_i}{\mu_i}\right) - (n_i - y_i) \log\left(\frac{n_i - y_i}{n_i - \mu_i}\right).$$

(d) Gamma,

$$\tfrac{1}{2} e_i^2 = \log\left(\frac{\mu_i}{y_i}\right) + \frac{y_i}{\mu_i} - 1.$$

The dispersion parameter in the gamma distribution does not appear in formula (d).

These are the most commonly required cases in practical data analysis. As pointed out in Ross (1982) the formula only applies if the likelihood with respect to μ has a single maximum when $\mu = y$. It is no doubt possible to postulate models for which this is not so, but in such cases maximum likelihood estimation often will be difficult to compute or interpret.

Curves showing e_i plotted against μ_i for typical values of y_i are shown in Fig. 2.8. The curvature reflects the asymmetry of the likelihood function caused by the constraints that the expectations be greater than zero (and less than n in the binomial case). The relationship between μ_i and e_i is in itself a transformation of the log-likelihood to quadratic form and therefore *stable* in the terminology of this chapter.

If the observations are not independent, but the covariance structure is known, it may be possible to define the log-likelihood in terms of the residuals of independent functions of the data. For example, in the simplest case of

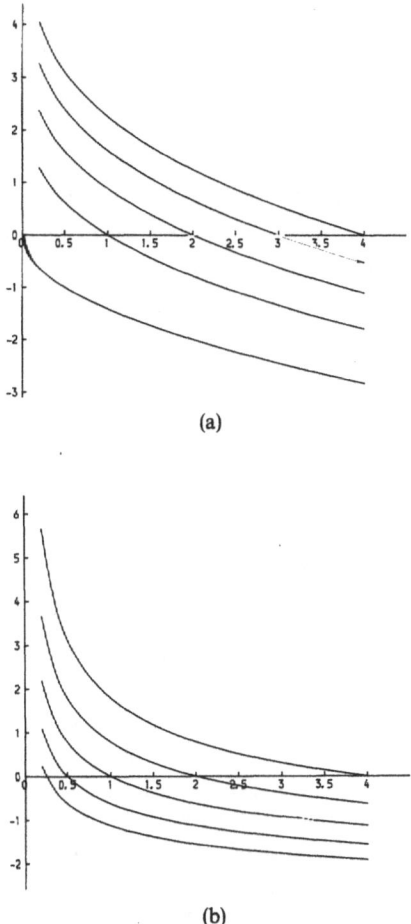

(a)

(b)

Fig. 2.8. Deviance residuals as functions μ given y. (a) Poisson distribution and (b) gamma distributions.

equally spaced observations in which the errors in y_i and y_{i+k} have correlation r^k, the quantities $y_1 - \mu_1$, $(y_2 - ry_1) - (\mu_2 - r\mu_1)$, $(y_3 - ry_2) - (\mu_3 - r\mu_2)$, etc., are independent, corresponding to derived observations $y_i - ry_{i-1}$ to which a model $\mu_i - r\mu_{i-1}$ is to be fitted.

2.4.1. Uses of Deviance Residuals

Besides the practical advantages of computing quantities that express, for nonnormal errors, the contribution to the residual deviance, and the computing methods available using these quantities (see Section 5.4.1) their theoretical importance is in simplifying the specification of the log-likelihood, independently of the parameter system used to specify μ_i.

A deviance residual is a function of parameters, and therefore a *potential parameter* in its own right. For example, if we select p generalized residuals from the set of n, such that the equations

$$\phi_j = e_j(\theta), \qquad j = 1, \ldots, p,$$

are of full rank, then $\hat{\theta}$ may be obtained, theoretically, from given values of $\hat{\phi}$. In the case of normal errors this is equivalent to the use of the expectations $\mu(\theta)$ as parameters, such as "stable ordinates" described in Section 2.2.3, since

$$e_i = \frac{y_i - \mu_i}{\sigma_i}$$

and y_i is fixed for the given data set. The maximum variance of e_i for a model with n parameters (a model of exact fit, with deviance 0, and each μ_i equal to the corresponding y_i), is unity, since the log-likelihood, expressed in terms of the e_i as parameters, is

$$L = -\sum \tfrac{1}{2} e_i^2,$$

and

$$\frac{\partial^2 L}{\partial e_i^2} = -1.$$

For any model with $p < n$ parameters it will be shown (Section 3.3) that the variance of e_i is less than 1, due to the pooling of information from neighboring observations, although for isolated observations the variance may be close to 1. In effect, we are saying that each fitted value is predicted by at least the observation itself, but that because of the continuity of the model, neighboring points provide extra replication. Hence deviance residuals, with zero expectation and bounded variance, satisfy some of the basic conditions for stable parameters.

A linear function of deviance residuals, $b'e$, also has zero expectation and finite variance not exceeding $b'b$. A nonlinear function, $B(e)$, has approximate variance (using the Taylor series expansion) $b'b$ where $b_i = dB/de_i$. This formula may be used to support the assertion that certain useful transformations of parameters are likely to be stable *a priori*.

The argument will be described geometrically in Section 4.2, in terms of the *solution locus*, or its generalization for nonnormal error models.

2.4.2. Examples of Stable Parameters Expressed in Terms of Deviance Residuals

(1) *Curve fitting*

We have used ordinates (expectations) as parameters to achieve approximately uncorrelated parameters and therefore quadratic likelihood functions.

For computational convenience equal spacing is preferred, but unequal spacing is usually necessary to minimize correlations.

When the chosen values of x correspond to data points, then the use of expectations, $f(x_i)$, as parameters, ϕ_i, is equivalent to the use of residuals,

$$e_i = y_i - f(x_i) = y_i - \phi_i \quad \text{or} \quad \phi_i = y_i - e_i.$$

If the chosen values are at other points between data values of x, say,

$$\phi = f(\lambda x_1 + (1-\lambda)x_2) \quad \text{where} \quad 0 < \lambda < 1,$$

then ϕ is a nonlinear function of the neighboring residuals

$$e_1 = y_1 - f(x_1), \qquad e_2 = y_2 - f(x_2),$$

and using, say Everett's formula for interpolation,

$$\phi = \lambda(y_1 - e_1) + (1-\lambda)(y_2 - e_2) + O(e_1^2, e_2^2).$$

For reasonably smooth curves the variance of expectations between data points should thus be bounded and fairly small, but for curves with very steep portions, or for extrapolated expectations, there is no guarantee that the variance will be small.

For curve fitting with nonnormal errors it is sufficient to use the above argument using the approximate error variance and the normal approximation, except when errors are large or expectations are close to the end of their permitted range. A more exact method is to use the appropriate transformations of observations and fitted values: for example, for gamma or lognormal errors to take logarithms, for Poisson variables to use the cube root transformation, and for binomial errors to use the incomplete beta transformation.

(2) *Fitting frequency distributions*
The connection between the suggested use of expected moments and deviance residuals as parameters in frequency distributions is less obvious than for curve fitting. Deviance residuals are obtained from the multinomial distribution for the observed frequencies n_i and the fitted frequencies Np_i where $N = \sum n_i$. Now we use the approximation

$$e = \frac{n_i - Np_i}{(Np_i(1 - p_i))^{1/2}},$$

and note that the fitted mean, $m = \sum t_i p_i$, where t_i is a typical value of x lying in the ith class interval (Section 1.1.4). Then

$$m = \sum t_i \left(\frac{n_i}{N} - e_i \left(\frac{p_i(1 - p_i)}{N} \right)^{1/2} \right)$$

$$= m^* - \sum e_i t_i \left(\frac{p_i(1 - p_i)}{N} \right)^{1/2},$$

showing that m is estimated by a provisional value m^* adjusted by a weighted sum of residuals.

The fitted variance is $s^2 = \sum u_i^2 p_i$, where $u_i + m$ is another value of x lying in the ith class interval. Again, s^2 in a linear function of p_i, and therefore of e_i, but weighted more unevenly than m, and depending mainly on the outer classes. This implies that s^2 (and s) is relatively more variable than m. By analogy, higher moments will be increasingly unstable as the weights involve higher powers of u_i, and concentrate the information almost entirely on the outer classes.

Percentiles, however, may be seen to be stable, because

$$\phi_k = F^{-1}(P_k) \qquad \text{or} \qquad F(\phi_k) = P_k,$$

where P_k is a chosen percentage such as 25% and ϕ_k is the corresponding value of x on the distribution function $F(x)$. Now, if the empirical percentile x_k lies in the jth interval, (x_{j-1}, x_j), then

$$\sum_{i=1}^{j-1} n_i \le NP_k \le \sum_{i=1}^{j} n_i,$$

whereas ϕ_k lies in the interval such that

$$\sum_{i=1}^{r-1} p_i \le P_k \le \sum_{i=1}^{r} p_i.$$

Now

$$\sum p_i = \frac{\sum n_i}{N} - \sum e_i \left(\frac{p_i(1 - p_i)}{N} \right)^{1/2},$$

and the sum of the residuals tends to be close to zero, especially for the middle range of the distribution, so that ϕ_k is likely to be close to x_k when the model is appropriate (i.e., the residuals e_i are not systematically grouped in sign). This idea is related to the property used in the Kolmogorov–Smirnov test, that the distribution functions of two samples from the same distribution should not differ by an excessive amount.

(3) *Multivariate models and contingency tables*
The idea that marginal expectations are more stable than individual expectations is reinforced by the observation that they are functions of a sum of several weighted residuals which tend, after fitting, to be of mixed sign. For if the signs are all the same the fit can always be improved by adjusting the appropriate parameter.

2.4.3. Data Dependence of Transformations

That deviance residuals form the basis for choice of parameter transformations implies that the best transformations are *data-dependent*. Only in the case of certain frequency distributions, where the shape of the sample is unaffected by design considerations, can theoretically stable parameters be deduced for all data sets.

This is why parameter transformation is essentially a computing activity. Rather than determining the exact form of certain algebraic or analytical expressions, algorithms are used which compute empirical statistics from the data and use these as working constants in transformations. The transformations may be changed dynamically as fitting proceeds, particularly if unexpected data-features cause difficulties at the first attempt.

2.4.4. Parameter Transformations and Measures of Curvature

Beale (1960) showed that the effects of nonlinearity in a model could be described partly in terms of the parametrization and partly in terms of the *intrinsic nonlinearity* due to curvature of the solution locus (see Section 4.3.1). Bates and Watts (1980) showed that for most practical examples the intrinsic nonlinearity, as measured by the curvature of the solution locus, is very small compared with the *parameter-effects curvature*.

While these findings are important, it is not necessary to compute curvature routinely if satisfactory parameter transformations may be found. Direct methods of computing confidence intervals (Section 5.5) express the effect of curvature in a more practical way than the measures proposed by Bates and Watts, which involve extensive computing of third-order quantities.

The main conclusion is that while nonlinearity cannot be completely eliminated by the transformation of parameters, the practical effects of intrinsic nonlinearity are very small, especially if there are sufficient data and the model is appropriate.

2.5. Similarity of Models

In Section 1.2.2 it was stated that it is not known how to compare models with the same number of parameters by statistical significance tests. However, the method of parameter transformation makes it possible to compare models in terms of the behavior of *analogous parameters*. Analogous parameters are parameters that have the same interpretation in different models. Simple examples are stable ordinates (fitted values used as parameters) in curve fitting, and expected moments in distribution fitting.

Consider the data of Table 2.1 (Section 2.1.1). The model fitted was a rectangular hyperbola through the origin with suggested parametrization

$$E(y) = \frac{\phi_1 \phi_2 x}{(2\phi_1 - \phi_2)x + 6(\phi_2 - \phi_1)},$$

taking the values ϕ_1 and ϕ_2 when $x = 3$ and 6, respectively. Alternative curves through the points $(3, \phi_1)$ and $(6, \phi_2)$ are the exponential,

$$E(y) = \frac{\phi_1^2}{2\phi_1 - \phi_2}\left(1 - \left(\frac{\phi_2 - \phi_1}{\phi_1}\right)^{x/3}\right),$$

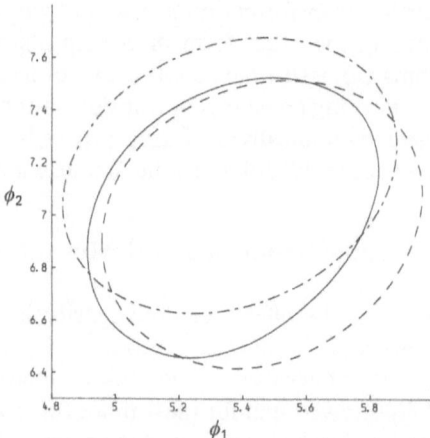

Fig. 2.9. Similar contours of residual sums of squares for three models fitted to the data of Table 2.1. The contour level relative to the optimum value is the same in each case. Continuous line—rectangular hyperbola; broken line—exponential; and short lines—quadratic.

and the quadratic

$$E(y) = \frac{x}{6}(4\phi_1 - \phi_2) - \frac{x^2}{18}(2\phi_1 - \phi_2).$$

Contours of log-likelihood for a common estimate of residual error variance are shown in Fig. 2.9. For the quadratic, a linear model, contours are perfect ellipses. The overall picture is very similar, but the detailed differences reflect differing behavior particularly as ϕ_1 approaches ϕ_2. Although the likelihood

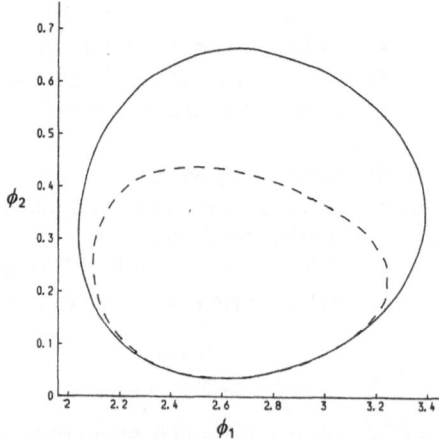

Fig. 2.10. Similar contours of log-likelihood for two models fitted to the data of Table 2.2. Full line—negative binomial distribution; broken line—Neyman Type A distribution.

contours are similar, the fitted optima differ. An explanation for the similarity is given in Section 4.3.1 where the solution-locus is discussed.

A similar comparison may be made for the data of Table 2.2 to compare the likelihood contours for the negative binomial distribution and the Neyman Type A distribution. The parameter system is based on the expected moments m and V, with second parameter

$$\phi_2 = \frac{V - m}{m^2}.$$

This is known to be approximately uncorrelated with $\phi_1 = m$ for the negative binomial, where it represents $1/k$. In the Neyman Type A distribution ϕ_2 represents $1/m_1$ and ϕ_1 represents $m_1 m_2$. Equivalent contours are shown in Fig. 2.10. Note that the contours tend to coincide as ϕ_2 tends to zero because both distributions become the Poisson distribution when $\phi_2 = 0$. The negative binomial contour is larger because the fit is better.

2.5.1. Use of the Similarity Concept

The importance of the similarity concept is that it allows *a priori* stable parameters to be chosen on the basis that a transformation that works well for one well-understood model should also work well for similar models. We understand orthogonal polynomial curves very thoroughly, and if a nonlinear model is graphically similar, over the range of the data, to a polynomial with the same number of parameters, then the approximate shape of the log-likelihood contours may be inferred.

Another use of similarity is in comparing models with the same expectations but different weights or error distributions. Although the scale of the log-likelihood function may change, its shape will depend mainly on the weights attached to various data observations. We may therefore be reasonably confident that a stable parametrization remains stable over a wide range of error distributions.

EXERCISES

1. Using any optimization algorithm of your choice, fit a nonlinear model to data using several different parametrizations. Always start from the same equivalent initial values. Record the number of iterations or function evaluations required to achieve convergence.
2. Use any suitable computer program to fit a nonlinear curve to some data. Using the dispersion matrix of parameter estimates, find the variance of the expectation and the slope at a number of interpolated and extrapolated values of x.

CHAPTER 3

Inference and Stable Transformations

3.1. Existence and Uniqueness of Solutions

Practical experience with fitting nonlinear models shows that for some data the model cannot be fitted at all, whereas for other data more than one solution may be found. Stable parameters enable us to describe in broad terms the characteristics of data sets likely to cause problems. For numerical analysts the answer may lie in seeking algorithms for global optimization, but for statisticians a deeper understanding of the problem is needed.

Let Y be the n-dimensional space of possible data sets y (a subspaceof R_n), and let D_θ be the p-dimensional domain of parameter sets θ. Then the likelihood function $L(\theta|y)$ may have:

(i) no proper optima in D_θ (although there must be a largest value on the boundary of D_θ);
(ii) a single, unique optimum $\hat{\theta}$; or
(iii) two or more local optima $\hat{\theta}^{(1)}, \hat{\theta}^{(2)}, \ldots$.

In the third case two or more optima may correspond to the same fitted values if the model consists of sums of similar terms, as in the double and triple exponentials discussed in Section 2.2.3(iv). In such cases the parameter domain should be restricted to avoid duplication of identical models, for example, by constraints such as $\theta_1 \geq \theta_2$.

Let $\phi(\theta)$ be a transformation from θ to ϕ so that the equivalent problem is to optimize $L(\phi|y)$ over the domain D_ϕ. Then there are several possibilities if the transformation is nonlinear.

(a) The solution is unique in D_ϕ but the equations $\phi(\theta) = \hat{\phi}$ cannot be solved for θ in D_ϕ. For example, if D_θ is the domain $\theta \geq 0$ and $\phi = \exp(-\theta)$, then we must have $0 < \hat{\phi} \leq 1$. If $\hat{\phi} > 1$ then $\hat{\theta} < 0$, and if $\hat{\phi} < 0$ then $\log \hat{\phi}$ is a complex number. This difficulty will be discussed further when we consider *pseudomodels* (Section 3.1.4).

(b) Equivalent multiple optima in D_θ transform into unique optima in D_ϕ. For example, if θ_1 and θ_2 are roots of the quadratic $t^2 + at + b = 0$, then the solutions (θ_1, θ_2) and (θ_2, θ_1) are equivalent.

(c) Distinct multiple optima in D_θ transform into more easily definable regions of D_ϕ so that unintended solutions $\hat{\theta}^{(2)}, \hat{\theta}^{(3)}, \dots$, etc., may be excluded from consideration. For example, the curve

$$y = \frac{P(x)}{Q(x)},$$

where $P(x)$ and $Q(x)$ are polynomials is discontinuous at roots of $Q(x) = 0$. To ensure that a continuous curve is fitted in the data range $x_a < x < x_b$, the parameters of $Q(x)$ may be transformed to define a region which excludes roots of $Q(x)$ between x_a and x_b.

The suggested procedure is to find suitable stable transformations $\phi(\theta)$ which involve parameters that estimate functions of the data. Detailed description of the parameter domain D_ϕ will then indicate the properties of data sets that show how many solutions exist in D_θ.

3.1.1. Exact Fits: As Many Parameters as Data Values

When the number of parameters equals the number of observations the solutions do not depend on the error distribution or the likelihood function. The simultaneous equations, $y_i = f_i(\theta)$, define solutions $\hat{\theta}$, and the parameters ϕ may be directly related to the observations y. Many important models may be discussed in terms of two parameters only, which allows the discussion to be expressed graphically in terms of charts showing the relationship between ϕ and θ for all possible y sets. The observations must of course be mutually independent.

The following cases will be discussed:

(i) Two-parameter models,

$$y_i = \phi_i, \qquad i = 1, 2.$$

(ii) Three-parameter models,

$$y_i = \theta_3 f_i(\theta_1, \theta_2),$$

$$\phi_i = \frac{y_i}{y_3}, \qquad i = 1, 2.$$

(iii) Four-parameter models,

$$y_i = \theta_4 + \theta_3 f_i(\theta_1, \theta_2),$$

$$\phi_i = \frac{y_{i+1} - y_1}{y_4 - y_1}, \qquad i = 1, 2,$$

or, if $y_i = \theta_4 f_i(\theta_1, \theta_2, \theta_3)$,

$$\phi_i = \frac{\log(y_{i+1}/y_1)}{\log(y_4/y_1)}, \qquad i = 1, 2.$$

In cases (ii) and (iii), parameters θ_3 and θ_4, representing linear scale and origin parameters, are eliminated from the discussion by division or subtraction of pairs of y values. The ϕ parameters represent shape, and are unaffected by changes of scale or origin in the data. All sets with given ϕ_1 and ϕ_2 exhibit the same problems.

One-parameter models, and two- or three-parameter models with linear scale and origin parameters, may be similarly treated. The graphical representation is simpler, a curve connecting $\hat{\phi}$ and $\hat{\theta}$ which shows whether solutions exist.

If three or more parameters affect shape, a series of charts may be required.

EXAMPLE 3.1 (The Two-Compartment Model). A process modeled by flow from one compartment to a second and then out of the system gives rise to the following equation for the total content of both compartments at time x relative to the total content at time 0

$$f(x, \lambda, \mu) = \frac{1}{\lambda - \mu}(\lambda e^{-\mu x} - \mu e^{-\lambda x}). \qquad (3.1.1)$$

This model is often discussed as an example of a difficult optimization problem with two parameters. It is effectively a three-parameter model with the scale parameter estimated by the value at $x = 0$. Let ϕ_1 and ϕ_2 be the observed values (or ratios) at $x = 1$ and $x = 2$, respectively. Then we must try to solve the simultaneous equations

$$\frac{1}{\lambda - \mu}(\lambda e^{-\mu} - \mu e^{-\lambda}) = \phi_1,$$

$$\frac{1}{\lambda - \mu}(\lambda e^{-2\mu} - \mu e^{-2\lambda}) = \phi_2. \qquad (3.1.2)$$

If (λ, μ) is a solution, so is (μ, λ) by symmetry. A chart of ϕ_1 and ϕ_2 for different values of $\lambda > 0$ shows that all solutions lie in a narrow region of (ϕ_1, ϕ_2) space (Fig. 3.1) between the curves A and B, where curve A corresponds to $\mu = \infty$ when $\phi_2 = \phi_1^2$, and curve B corresponds to the limit as μ tends to λ,

$$\phi_1 = (1 + \lambda)e^{-\lambda}, \qquad \phi_2 = (1 + 2\lambda)e^{-2\lambda}.$$

The solution region, D_λ is quite narrow. For example, if $\phi_1 = 0.4$ then ϕ_2 must lie between 0.089 and 0.160, and if $\phi_1 = 0.8$ then ϕ_2 must lie between 0.510 and 0.640. To solve the equations (3.1.2) for λ and μ a slowly converging iterative scheme is as follows, with the convention $\lambda \le \mu$:

$$e^{-\lambda} = \frac{(\phi_2 - \phi_1 e^{-\mu})}{(\phi_1 - e^{-\mu})},$$

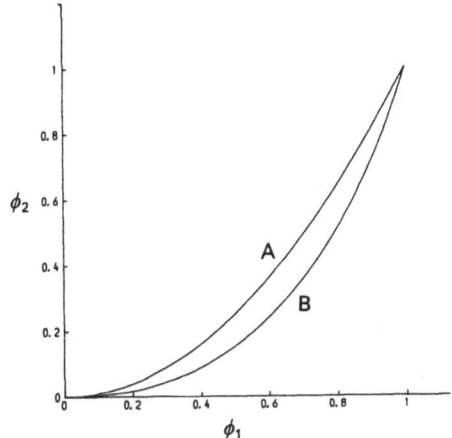

Fig. 3.1. Solution region for the two-compartment model. Real solutions of the transformation from (ϕ_1, ϕ_2) to (λ, μ) exist only between the two curves A and B (see text).

$$\mu = \frac{\lambda(\phi_1 - e^{-\mu})}{(\phi_1 - e^{-\lambda})},$$ (3.1.3)

starting with $\mu = \infty$, $e^{-\lambda} = \phi_2/\phi_1$. Then new values of μ and λ are evaluated alternately until convergence. Outside the solution region the process diverges.

EXAMPLE 3.2 (The Rectangular Hyperbola). The model

$$E(y) = \frac{\theta_2 x}{x + \theta_1}$$ (3.1.4)

was discussed in Section 2.1.4. It represents a rectangular hyperbola that passes through the origin and has a horizontal asymptote $y = \theta_2$ and a vertical asymptote $x = \theta_1$. For some appropriate scaling of x we can write

$$\phi_1 = \frac{\theta_2}{1 + \theta_1}, \qquad \phi_2 = \frac{2\theta_2}{2 + \theta_1},$$

whence

$$\theta_1 = \frac{2(\phi_2 - \phi_1)}{2\phi_1 - \phi_2}, \qquad \theta_2 = \frac{\phi_1 \phi_2}{2\phi_1 - \phi_2}.$$ (3.1.5)

In applications, ϕ_1 and ϕ_2 are always positive, but acceptable solutions are confined to the triangular region between the lines $\phi_2 = \phi_1$ and $\phi_2 = 2\phi_1$. Outside this region θ_1 is negative, producing a discontinuous curve with vertical asymptote at some positive value of x.

EXAMPLE 3.3 (The Generalized Hyperbola). The model

$$E(y) = \alpha + \frac{\beta}{(1 + \gamma x)^{-\delta}}$$ (3.1.6)

(equation (2.2.23)) was proposed by Turner (1962) as a flexible, all-purpose model for monotone increasing data, it includes the following special cases:

$$\delta \to 0, \qquad E(y) = \alpha + \beta \log (1 + \gamma' x),$$

$$\delta = 1, \qquad E(y) = \alpha + \frac{\beta}{1 + \gamma x},$$

$$\delta \to \infty, \qquad E(y) = \alpha + \beta \exp (\gamma' x),$$

$$\delta = -1, \qquad E(y) = \alpha' + \beta x.$$

In practice it has seldom been used, and the reason why becomes clear when we study the solution region for four equally spaced data values at $x = 0$, 1, 2, and 3.

We eliminate α and β by computing

$$\phi_1 = \frac{y_1 - y_0}{y_3 - y_0}, \qquad \phi_2 = \frac{y_2 - y_0}{y_3 - y_0}.$$

For fixed δ the limit as γ tends to 0 is a straight line, so that curves relating ϕ_1 and ϕ_2 pass through (1/3, 2/3). For positive δ as γ tends to infinity ϕ_1 and ϕ_2 both tend to 1, but for negative δ they both tend to 0, so we must also consider the case where γ is negative. Provided $(1 + \gamma x)$ is positive the model is defined. Although hardly satisfactory as an explanation of real phenomena, the limiting cases are as follows:

$$\text{if} \quad \delta < -1, \qquad \phi_1 = 1 - (2/3)^{-\delta}, \qquad \phi_2 = 1 - (1/3)^{-\delta},$$

$$\text{if} \quad -1 < \delta < 0, \qquad \phi_1 = (1/3)^{-\delta}, \qquad \phi_2 = (2/3)^{-\delta}.$$

The solution region is illustrated in Fig. 3.2. Curve A is the limit for the case

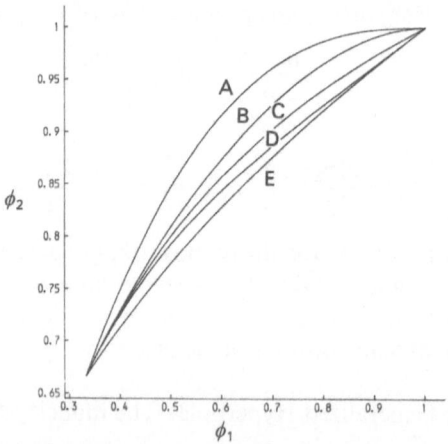

Fig. 3.2. Solution region for the generalized hyperbola. Real solutions of the transformation from (ϕ_1, ϕ_2) to (γ, δ) exist only between curves A and E. For explanation of the curves, see text

$-1 < \delta < 0$. Curve B corresponds to $\delta = 0$ (logarithmic models), curve C to $\delta = 1$ (rectangular hyperbolas), and curve D to $\delta = \infty$ (exponentials). Curve E is the limit for the case $\delta < -1$.

The inner region between curves B and D gives unique interpretable solutions with $\gamma, \delta < 0$. For γ and δ of opposite sign a similar diagram can be produced between the origin $(0, 0)$ and the "straight line" point $(1/3, 2/3)$, obtained simply by reversing the sign of x or the order of the observations.

The following table demonstrates the narrowness of the acceptable region:

ϕ_1	ϕ_2 for curves				
	A	B	C	D	E
0.4	0.713	0.726	0.727	0.729	0.749
0.5	0.774	0.793	0.800	0.809	0.847
0.8	0.828	0.844	0.857	0.875	0.917
0.7	0.877	0.887	0.903	0.927	0.962
0.8	0.921	0.926	0.941	0.966	0.987
0.9	0.962	0.963	0.973	0.991	0.998

The best iterative solution for γ and δ given ϕ_1 and ϕ_2 depends on which curve (A to B) is nearest the given values. Near curve D γ^δ tends to a constant and the exponential solution provides a starting value for a solution in terms of γ^δ and γ,

$$\phi_2 = \frac{1 + e^{-\gamma^\delta}}{1 + e^{-\gamma^\delta} + e^{-2\gamma^\delta}},$$

whereas below curve C it is simplest to start from the hyperbola solution,

$$\delta = 1, \qquad \gamma = \frac{3\phi_1 - 1}{3(1 - \phi_1)},$$

and solve for γ and δ by Newton's method of successive approximation using differences rather than derivatives.

EXAMPLE 3.4 (The Modified Exponential). An example of a model in which there may be more than one distinct solution is the modified exponential

$$E(y) = \alpha + (\beta + \gamma x) \exp(-kx), \qquad (3.1.7)$$

which arises as the limiting case of the sums of exponentials when the two nonlinear parameters become equal. This curve has a maximum or minimum at $x = (\gamma - \beta k)/\gamma k$, and is not monotone. The range of possible values of the ratios of differences of ordinates is unlimited, but the range of solutions is again restricted.

The most convenient treatment for algebraic analysis is to define

$$\phi_1 = \frac{y_2 - y_1}{y_1 - y_0}, \qquad \phi_2 = \frac{y_3 - y_2}{y_1 - y_0}.$$

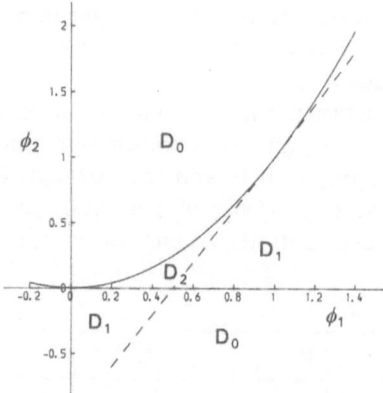

Fig. 3.3. Solution regions for the modified exponential. D_0, no acceptable solutions; D_1, one acceptable solution, one unacceptable solution; D_2, two acceptable solutions.

Then $r = e^{-k}$ is a root of

$$r^2 - 2r\phi_1 + \phi_2 = 0. \tag{3.1.8}$$

Now we generally require $0 < r < 1$ for a curve with an asymptote as x tends to infinity ($k > 0$). There are therefore several regions depending on the roots of (3.1.8), as illustrated in Fig. 3.3.

Above the parabola $\phi_2 = \phi_1^2$ there are no real solutions (Region D_0). Between the lines $\phi_2 = 0$ and $\phi_2 = 2\phi_1 - 1$ (tangents to the parabola at $\phi_1 = 0$ and $\phi_1 = 1$, respectively), there is one acceptable solution ($0 < r < 1$), and one unacceptable solution $r > 1$ or $r < 0$ (Region D_1). Between the parabola and the two tangents there are two acceptable solutions (Region D_2). Below the two lines there are no acceptable solutions (Region D_0). The shapes of curves for the two acceptable solutions are quite close within the data range, but diverge outside it as illustrated by the following table for the case $\phi_1 = 0.5$, $\phi_2 = 0.29$, for the two solutions:

Curve 1: $y = 1.975 - (1.875 - 0.4167x)0.6^x$.
Curve 2: $y = 1.944 - (1.944 + 0.4167x)0.4^x$.

x	y (Curve 1)	y (Curve 2)
-2	-5.648	-5.000
-1	-1.440	-1.875
0	0.000	0.000
0.5	0.584	0.583
1	1.000	1.000
1.5	1.294	1.294
2	1.500	1.500
2.5	1.643	1.642
3	1.871	1.852
4	1.889	1.944

Curve 1 has a maximum at $x = 4.5$ while curve 2 has a minimum at $x = -4.667$. Effectively, the model is being used to fit data over too short a range, which indicates that more parameters are being used than are justified by the data.

3.1.2. Nonexact Fits: More Data Values than Parameters

When studying the most usual case, when the number of samples exceeds the number of parameters, little use can be made of the geometric approach. We wish to subdivide n-dimensional data space Y into regions characterized by the number of acceptable maximum likelihood solutions. The error distribution now becomes relevant; except for data sets corresponding to the locus of exact fits (known as the *solution locus* (see Section 4.3.1)) the estimates change with the error distribution, and critical boundaries change also.

Two-dimensional study is only possible for the following cases:

(i) Two observations and one parameter, $y_1 = f_1(\theta_1)$, $y_2 = f_2(\theta_1)$.
(ii) Three observations and two parameters, $y_i = \theta_2 f_i(\theta_1)$.
(iii) Four observations and three parameters, $y_i = \theta_3 + \theta_2 f_i(\theta_1)$.

However trivial these may seem, the examples are instructive. The technique is to plot the solution locus (exact fits) for the data, or ratios of data, or ratios of differences of data, respectively, for the three models, as parametrized by θ_1. This is very easy to do. Now, given the likelihood function $L(y|\theta)$, obtain the normal equations $\partial L/\partial \theta = 0$ and eliminate θ_2 and θ_3 if possible to obtain a relationship between θ_1 and the data values. For each value of θ_1 a locus (a straight line for weighted normal errors) may be drawn. Points on this locus form a set of data values with a common MLE, although parts of the locus may correspond to minima rather than maxima. Such loci may be called *common estimate loci* (see Section 4.3.1). The number of solutions is therefore the number of common estimate loci passing through a given point corresponding to the data set.

EXAMPLE 3.5 (Exponential Decay). Let a process measured at times $x = 1$ and $x = 5$ decay according to the law

$$E(y) = \theta^x, \quad 0 < y < 1, \quad \theta < 1,$$

where y is the proportion of the initial measurement at $x = 0$. The solution locus is $y_1 = \theta$, $y_5 = \theta^5$, or $y_5 = y_1^5$. Assuming independently distributed normal errors with equal variance the normal equation is

$$y_1 + 5\theta^4 y_5 = \theta + 5\theta^9, \qquad (3.1.9)$$

which represents a family of normals (lines perpendicular) to the solution locus (with apologies for the three different uses of the word "normal"), the common estimate loci. These loci are not parallel and therefore intersect, and Fig. 3.4 shows the typical cusped curve which is the envelope of the system. In

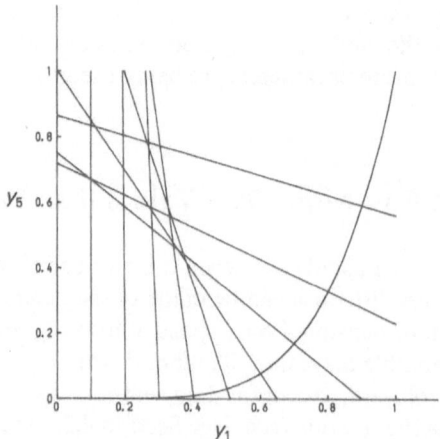

Fig. 3.4. Solution locus and common estimate loci for the exponential curve through two points (y_1, y_5) with normal errors. Curve $y_5 = y_1^5$ is the solution locus, and normal lines are drawn for $\theta = 0.1\ (0.1)0.9$ (see text).

elementary geometry the envelope of normals to a curve is known as the *evolute*, and it is the locus of centers of curvature. Cusps correspond to points on the original curve of greatest curvature. The region within the cusped locus corresponds to data for which there are two solutions. Three loci pass through each point: one corresponds to a maximum sum of squares and two to a minimum. Elsewhere there is a single solution within the square, $0 < y_1$, $y_5 < 1$.

This example indicates that multiple solutions are more likely when the fit is poor. This does not mean that good fits are always unique, as we have seen in the exact examples above, where parts of the solution locus corresponded to more than one solution. But for simpler nonlinear models it can be shown that where the solution locus is of low curvature, multiple solutions indicate a very poor fit.

Now suppose that y is a binomial variable with equal sample size for y_1 and y_5. Then the log-likelihood is proportional to

$$y_1 \log \theta + (1 - y_1) \log(1 - \theta) + y_5 \log \theta^5 + (1 - y_5) \log(1 - \theta^5)$$

and the normal equation is

$$y_1(1 - \theta^5) + 5y_5(1 - \theta) = \theta(1 - \theta^5) + 5\theta^5(1 - \theta). \qquad (3.1.10)$$

This represents a system of loci which do not intersect within the unit square, as illustrated in Fig. 3.5. The conclusion is that for binomially distributed variables the solution is unique for all possible data. This illustrates the advantage of postulating an appropriate error distribution.

EXAMPLE 3.6 (Exponential Curves Through Four Points). The model $E(y) = \alpha + \beta e^{-kx} = \alpha + \beta\rho^x$ fitted to four equally spaced points by least squares may

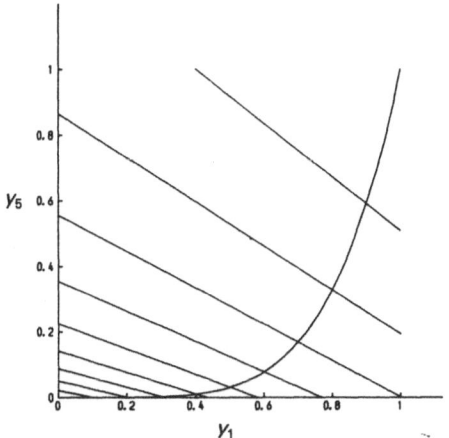

Fig. 3.5. Solution locus and common estimate loci for the exponential curve through two points (y_1, y_5) with binomial distribution. Curve $y_5 = y_1^5$ is the solution locus, and the straight lines correspond to estimates of $\theta = 0.1\ (0.1)0.9$ (see Fig. 3.4).

be illustrated in terms of the data ratios

$$z_1 = \frac{y_1 - y_0}{y_3 - y_0}, \qquad z_2 = \frac{y_2 - y_0}{y_3 - y_0}.$$

The normal equations are for estimates (a, b, r)

$$
\begin{aligned}
na & + b \sum r & = \sum y, \\
a \sum r^x & + b \sum r^{2x} & = \sum yr^x, \\
a \sum xr^{x-1} & + b \sum xr^{2x-1} & = \sum xyr^{x-1},
\end{aligned}
$$

and eliminating a and b we obtain the equation

$$\sum (y - \bar{y})r^{x-1}\left(\sum r^x - \frac{(\sum r^x)^2}{n} \right) = \sum (y - \bar{y})r^x \left(\sum xr^{2x-1} - \frac{\sum r^x \sum xr^{x-1}}{n} \right)$$

for $x = 0, 1, 2$, and 3. This equation is not changed by subtracting y_0 from each y and dividing by $y_3 - y_0$. Expanding the sums and dividing by powers of $1 - r$ the equation becomes

$$(2 + 4r + 3r^2 + 2r^3 - r^4)z_1 + (-1 + 2r + 3r^2 + 4r^3 + 2r^4)z_2$$

$$= 1 + 4r + 3r^2 + 2r^3. \tag{3.1.11}$$

This parameter system of lines is illustrated in Fig. 3.6. Exact solutions lie on the curve

$$z_1 = \frac{1}{1 + r + r^2}, \qquad z_2 = \frac{1 + r}{1 + r + r^2},$$

and solutions are unique in this region of data space (where the fit is good).

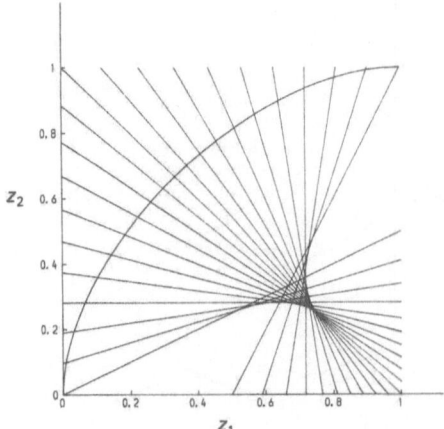

Fig. 3.6. Common estimate loci for three-parameter exponential curves fitted to four points with normal errors. Data standardized by fixing two points give identical diagrams in terms of the nonlinear parameter estimate r. Lines are solutions of (3.1.11) for $r = 0$ (0.1)1 and $r^{-1} = 0$ (0.1)1. The curve corresponds to exact fits (see text).

Multiple solutions occur in the small triangular area around $z_1 = 2/3$, $z_2 = 1/3$ where the fit is poor. Below the critical curve there are no solutions either for $0 < r < 1$ or $1 < r < \infty$, because the values of r parametrizing the line are local maxima of the residual sum of squares, rather than minima.

The set of lines (3.1.11) may be viewed as a ruled surface in three dimensions (z_1, z_2, and r), resembling the "breaking wave" of Catastrophe Theory. Small changes in the data (z_1 and z_2) can bring the solution status from uniqueness to multiplicity where the vertical projection of the surface overlaps itself. For any given point (z_1, z_2) the residual sum of squares may be plotted as a function of r, or better of $1/(1 + r) = h$, where h ranges between 0 and 1. An example of the gradual transition from unique to multiple solutions is given in Fig. 3.7.

In practice, it is also important that r be either greater or less than 1, because the model is applied either to data where an asymptote is expected ($r < 1$) or to nonasymptotic growth processes ($r > 1$). Therefore the line $z_1 + z_2 = 1$ divides the data space into two regions dominated by $r < 1$ or $r > 1$.

The region of the solutions is mainly bounded by the lines $z_2 = 2z_1 - 1$ and $z_2 = \frac{1}{2}z_1$. This shows that for there to be a solution with $0 < r$ the second point must not exceed the mean of the third and fourth. Thus, although the strict monotone condition for three points (exact fits) is relaxed, only slight extensions are possible.

EXAMPLE 3.7 (Rectangular Hyperbolas Through Four Points). The rectangular hyperbola may be analyzed in a very similar fashion to the preceding example, merely by substituting for r^x the expression $1/(1 + dx)$ and transforming to the parameter $h = (1 + 3d)/(2 + 3d)$ when, after cancellation of

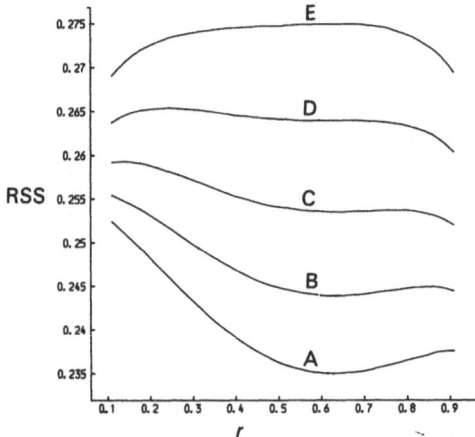

Fig. 3.7. Residual sum of squares RSS(L) for a series of data sets from Fig. 3.6 corresponding to $r = 0.7$. RSS(L) plotted as a function of $h = 1/(1 + r)$. Curves show: A, unique minimum; B, two minima; C, points of inflexion but no local minimum; D and E, local maxima.

common factors, the common estimate loci are

$$(-2 + 13h - 12h^2 + 13h^3 - 5h^4)z_1 + (7 - 8h - 3h^2 + 7h^3 - 5h^4)z_2$$
$$= 13h - 24h^2 + 21h^3 - 5h^4. \tag{3.1.12}$$

These loci are shown in Fig. 3.8. The region of multiple solutions is slightly different in shape from that in Fig. 3.6, but overall the regions are very similar, reflecting the similarity in shape of the hyperbola and the exponential curves.

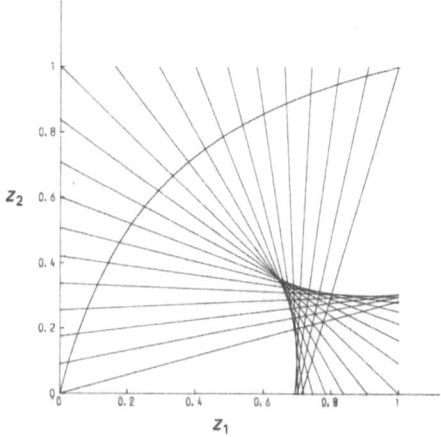

Fig. 3.8. Common estimate loci for three-parameter rectangular hyperbolas fitted to four points with normal errors. Data standardized by fixing two points give identical diagrams in terms of the nonlinear parameter estimate h. Lines are solutions of (3.1.12) for $h = 0\,(0.05)1$. The curve corresponds to exact fits (see Fig. 3.6).

To extend these investigations to larger samples or to models with more parameters would be very complicated. Geometrically, the common estimate loci become linear subspaces (planes or primes), but how they intersect can only be determined by looking at two-dimensional cross sections.

An alternative approach is to generate random data points and to plot the complete likelihood function for each, where this is possible. A study of the exponential model fitted to six points showed that the pattern of cases with unique solutions, multiple solutions, and nonexistent solutions was broadly similar to that for four points.

A third approach is to plot parameter loci (see Section 4.4.5) showing the relationships between parameters for each observation or group of observations separately. These loci in parameter space should pass close to a single point if the fit is good and the estimate is unique. If the loci diverge no estimate exists, but if they intersect in several places multiple solutions may exist.

3.1.3. General Conclusions on Uniqueness of Estimates

While there can be no certainty in every case that the correct number of estimates is identified the following guidelines may be used:

(a) When the fit is "good" and the design "adequate" the solution is probably unique in stable parameter space, any ambiguities arising from the inverse transformation to defining parameters.
(b) When the fit is "poor", alternative equally inadequate estimates may exist.
(c) When the fit is good but the design is inadequate the transformation to acceptable defining parameters may be impossible, so that no solution exists. By design is meant the values of controllable variables that affect the relationship between data and model.
(d) If the error distribution is inappropriate, nonuniqueness of estimates is more likely.

3.1.4. Pseudomodels: Extending the Solution Region

When the stable parameter solution cannot be transformed back to a solution in terms of defining parameters, it is sometimes possible to introduce another model which has the same relationship with stable parameters but which applies to the complementary region of stable parameter space.

An obvious example is the double exponential model

$$E(y) = \alpha + \beta_1 \exp(-k_1 x) + \beta_2 \exp(-k_2 x), \qquad (3.1.13)$$

which satisfies the second-order linear differential equation

$$y'' + sy' + ty = C.$$

Depending on the sign of $(s^2 - 4t)$ the exponential terms are real or complex, and in the latter case the solution is written in terms of new parameters γ and k

$$E(y) = \alpha + \beta_1 \exp(-kx) \cos \gamma x + \beta_2 \exp(-kx) \sin \gamma x, \qquad (3.1.14)$$

or, if $s^2 = 4t$,

$$E(y) = \alpha + \beta(1 + \gamma x)\exp(-kx). \tag{3.1.15}$$

Treating s and t as stable parameters there is clearly a smooth transition in shape between models (3.1.13) and (3.1.14) as we move across the critical parabola, $s^2 = 4t$. Data with a slight tendency to oscillate might well have a solution in terms of model (3.1.14) but not in terms of model (3.1.13), which would not be realized from a study of k_1 and k_2 alone. Models such as (3.1.14) may be called *pseudomodels*.

Let the parameters θ of the original model $M(\theta)$ be transformed to parameters $\phi(\theta)$, and let D_ϕ be the region of ϕ space corresponding to θ space. A *pseudomodel*, $M^*(\psi)$, is a model for which there is a transformation $\phi(\psi)$ from ψ space into a region D_ϕ^* sharing a common boundary with D_ϕ but not overlapping. The purpose of pseudomodels is to extend the region D_ϕ so that the likelihood function, with its optima and contours defining confidence regions, may be computed in stable parameter space. Only at the interpretation stage does it become relevant whether the model is the required model or one of its associated pseudomodels.

The following table lists some examples where pseudomodels are likely to be useful.

Model	Stable parameters	Pseudomodels
1. Sums of exponentials	Differential equation coefficients	Damped oscillations
		Differences of exponentials
		Hyperbolic functions
2. Ratios of polynomials (continuous solutions)	Selected ordinates	Discontinuous solutions, with vertical asymptotes between data observations
3. Logistic curve	Selected ordinates	Inverse exponentials with vertical asymptotes
4. Negative binomial distribution parameters $m, k > 0$	Expected mean and variance	Positive binomial distribution, parameters $m, n = -k$.
5. Probit regression line with estimated control mortality > 0		Estimated control morality < 0.
6. Power law models, various (positive arguments only)		Treat powers of negative arguments $(-a)^b$ as $a^b \cos \pi b$
		This approach is not scale-invariant
		Treat $(-a)^b$ as 0.
7. Mixture of normal distributions	Expected moments	Allow negative components provided the expected frequency is always positive.

If pseudomodels are useful in allowing estimates to be obtained for a wider range of data sets, they are even more useful in allowing likelihood-based confidence regions to be described as complete ellipsoidal regions rather than as incomplete hyperboloidal regions. For a given sample size and critical contour we can imagine a sequence of data sets in which the following conditions hold:

(1) Estimate and contour entirely within D_ϕ.
(2) Estimate within D_ϕ, contour partly outside.
(3) Estimate outside D_ϕ, contour partly inside.
(4) Estimate and contour entirely outside D_ϕ.

EXAMPLE 3.8 (The Rectangular Hyperbola). In Example 3.2 it was shown that for the model (3.1.4) the transformation from ϕ to θ (3.1.5) was only acceptable within a triangular region between the lines $\phi_2 = \phi_1$ and $\phi_2 = 2\phi_1$. In Fig. 3.9 confidence regions determined by the constraint RSS $(\phi) < 2$ are illustrated for the following sets of exactly fitted data:

Set	$\hat{\phi}_1$	$\hat{\phi}_2$	θ_1	θ_2	$x = 0.5$	$x = 1$	$x = 1.5$	$x = 2$	$x = 2.5$
1	5	8	3	20	2.857	5.000	6.667	8.000	9.091
2	5	9	8	45	2.647	5.000	7.105	9.000	10.714
3	5	11	-12	-55	2.391	5.000	7.857	11.000	14.474
4	5	12	-7	-30	2.308	5.000	8.182	12.000	16.667

The four sets correspond to the four conditions above. Clearly, the classification depends on the critical value chosen, and the size increases with the variability of the data and with the choice of significance level.

For more complex models, where the acceptable region D_ϕ is relatively narrow, case 2 is likely to be the norm rather than the exception. For example, in the two-compartment model (Example 3.1) data must be very precise or highly replicated for an acceptable confidence contour to be produced. In many cases the contour may cross more than one critical boundary.

For the negative binomial distribution, where the probability of observing r events is defined as

$$P_r = \frac{\Gamma(r + k)}{r! \, \Gamma(k)} \left(\frac{m}{m + k} \right)^r \left(1 + \frac{m}{k} \right)^{-k}, \qquad r = 0, 1, \ldots, \qquad (3.1.16)$$

maximum likelihood estimates of m and k are independent, but k can tend to infinity. Large negative values of k are interpretable as the index $n = |k|$ of a positive binomial distribution. Estimates and confidence intervals are therefore better expressed in terms of parameters such as $1/k$, where as k tends to infinity there is a smooth transition from positive to negative values, with the special case $1/k = 0$ corresponding to the Poisson distribution. For example, when $\hat{m} = 1$ and $\hat{k} = 3$ for samples of 100, the 95% confidence limits for k

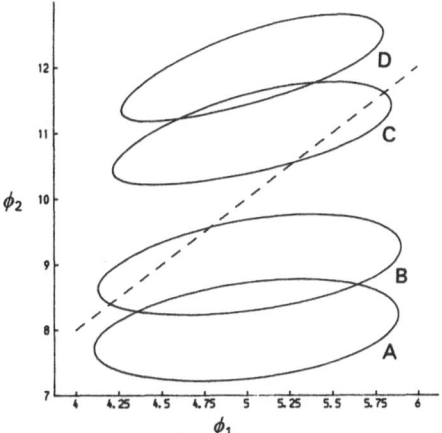

Fig. 3.9. Contours of residual sums of squares showing four types of solution: A, contour entirely within valid region; B, valid estimate but contour partly invalid; C, invalid estimate but contour partly valid; D, contour entirely outside valid region. For details see text.

(given m) are (1.16, 140) but for $1/k$ are (0.007, 0.861). When $\hat{k} = 5$, the limits for $1/k$ are $(-0.079, 0.662)$, transforming to the discontinuous interval for k, $k > 1.51$, or $k < -12.6$.

In practice, consideration of pseudomodels is preferable to the customary procedure (e.g., Willson et al., 1984) of discarding data sets that do not give acceptable θ parameters, and of computing sampling statistics of the remainder to obtain estimates of bias and precision. In the case of the negative binomial distribution the occurrence of extremely large values of k is possible, and all its sample moments are theoretically infinite.

The occurrence of a pseudomodel estimate, in isolation, may prevent further use of the ordinary defining model. But viewed as a sampling extreme among a group of similar sets of data, joint estimates of the defining parameters may be possible, and the ability to combine stable parameters may resolve the difficulty. For example, the two-compartment model may well be "true" but because of sampling error most data sets produce no acceptable parameter estimates. But the joint estimate over several sets may be acceptable.

EXAMPLE 3.9 (Fitting Exponential Curves to Several Sets of Data). Let the model, $E(y) = \alpha + \beta \exp(-kx)$, be fitted to the data in the following table:

Set	$x = 0$	$x = 1$	$x = 2$	$x = 3$	$\hat{\alpha}$	$\hat{\beta}$	$\hat{k}$
1	7	15	18	19	19.6	-12.65	1.007
2	11	14	16	21		k negative	
3	8	17	14	16		no fit possible, k complex	

Fitting a common value of k gives the following results:

Set	$\hat{\alpha}$	$\hat{\beta}$	$\hat{k}$
1	20.05	−12.97	0.905
2	19.11	− 8.84	0.905
3	17.04	− 8.06	0.905

Fitting a common of β and k gives:

Set	$\hat{\alpha}$	$\hat{\beta}$	$\hat{k}$
1	18.74	−9.89	0.928
2	19.49	−9.89	0.928
3	17.74	−9.89	0.928

Fitting a common curve of all data gives $\hat{\alpha} = 18.66$, $\hat{\beta} = -9.89$, $\hat{k} = 0.928$.

Treating the first three expectations as stable parameters, we can fit pseudo-models to data sets 2 and 3. Set 2 is a positive exponential and set 3 is a damped oscillation. The fitted values are as follows:

Set	$\hat{\phi}_1$	$\hat{\phi}_2$	$\hat{\phi}_3$
1	7.00	15.02	17.95
2	11.25	13.40	16.47
3	8.08	17.39	14.22
Mean	8.78	15.27	16.22
Common curve	8.76	14.75	17.11

The mean parameters are not very different from those obtained by fitting the model to the mean data values, but are sufficiently different to show that bias exists in nonlinear estimation unless we choose the exact unbiased transformation for that set of data, something which cannot be done *a priori*.

3.2. Inferences on Functions of Parameters

Any particular function of parameters $g(\theta)$ may be studied by treating it as a parameter in transformed space, replacing one of the existing parameters. However, for routine analysis this is not a useful procedure; many functions are required, and the inverse transformations may be very inconvenient to compute.

The problem is to obtain variances and confidence limits for the required functions, and to estimate the covariances of pairs of functions. The most important applications concern:

(a) the original parameters as functions of stable parameters;
(b) functions directly expressed in the form $g(\theta)$; and
(c) functions indirectly expressed in the form $\phi(g,\theta) = 0$.

For example, using a set of stable ordinates as parameters for a fitted curve we may be interested in obtaining variances and confidence limits for the defining parameters, for fitted points on the curve, and for inverse interpolates (solving for x given y).

Given parameter estimates $\hat{\theta}$ with dispersion matrix V, then the variance of a linear function of $\hat{\theta}$, $\mathbf{a}'\hat{\theta}$ is $\mathbf{a}'V\mathbf{a}$, and the covariance of $\mathbf{a}'\hat{\theta}$ and $\mathbf{b}'\hat{\theta}$ is $\mathbf{a}'V\mathbf{b}$. For nonlinear functions $g(\theta)$ the vector $\mathbf{a}$ is the vector of first partial derivatives $g_j'(\theta)$. For implicit functions of the form $\phi(g, \theta) = 0$ we can use the relationship

$$\frac{\partial \phi}{\partial \theta} + \frac{\partial \theta}{\partial g} g' = 0$$

and solve for g'.

EXAMPLE 3.10 (Rectangular Hyperbola (see Table 2.10)). The estimated dispersion matrix V of the fitted parameters in the model

$$E(y) = \frac{\theta_2 x}{x + \theta_1}$$

is as follows:

$$V = \begin{pmatrix} 0.4288 & 0.6148 \\ 0.6148 & 0.9589 \end{pmatrix} \quad \text{where} \quad \hat{\theta} = \begin{pmatrix} 2.549 \\ 9.956 \end{pmatrix}.$$

The expected value for $x = 3$ is $3\theta_2/(3 + \theta_1) = 5.383$ whence

$$g'(\theta) = \begin{pmatrix} -0.959 \\ 0.541 \end{pmatrix}$$

and the variance is 0.037. The value of x for which $y = 8$ is 10.422. Assuming we cannot easily solve for x algebraically, but can evaluate

$$\frac{\partial y}{\partial x} = 0.1508, \qquad \frac{\partial y}{\partial \theta_1} = -0.6167, \qquad \frac{\partial y}{\partial \theta_2} = 0.8035,$$

then

$$g'(\theta) = \begin{pmatrix} +4.089 \\ -5.328 \end{pmatrix}$$

and the variance is 7.602.

3.2.1. Approximate Confidence Intervals
for Functions of Parameters

Likelihood-based confidence regions (Section 1.2.2) for the parameters θ are defined in terms of critical contours of the log-likelihood function, $L(\theta)$, being regions in which θ satisfies

$$L(\theta) \leq L_c.$$

Confidence intervals for each parameter are determined by the maximum and minimum values of θ on the critical contour, assuming that θ is continuous and the critical contour is closed. A computational procedure for finding these values for θ_j is to optimize $L(\theta)$ with respect to all parameters except θ_j, and by interpolation in a sequence of optimum values $L^*(\theta_j)$ to solve

$$L^*(\theta_j) = L_c$$

for θ_j to obtain its upper and lower limits. If we assume that $L^*(\theta_j)$ is approximately of the form $L^*(\hat{\theta}) + k(\theta_j - \hat{\theta}_j)^2$, interpolation with respect to the square root of

$$L^*(\theta_j) - L^*(\hat{\theta}_j)$$

is approximately linear, which provides a reasonably fast algorithm when the limits are not too asymmetric (see also Section 5.5.1).

Confidence limits for a function $g(\theta)$ can be computed exactly using the above algorithm applied to a parameter transformation in which $g(\theta)$ replaces one of the parameters θ_j. Functions of a single parameter, of course, do not need this treatment provided there is a monotone relationship between θ and $g(\theta)$ over the interval. The limits are simply $g(\theta_{\min})$ and $g(\theta_{\max})$, although their order may be reversed.

For general functions which are single-valued and very approximately linear over the region within the critical contour we expect the maximum and minimum to occur on the contour itself. In the simplest case where the contour is exactly the ellipsoid

$$L(\theta) = L(\hat{\theta}) + \tfrac{1}{2}(\theta - \hat{\theta})' \, V^{-1}(\theta - \hat{\theta}) = L_c, \tag{3.2.1}$$

then the maximum or minimum may be found by Lagrange's method of undetermined multipliers as follows (Ross, 1978). We minimize

$$g(\theta) - \lambda(L(\theta) - L(\hat{\theta})) \tag{3.2.2}$$

with respect to λ and θ to obtain the following equations:

$$\frac{\partial g}{\partial \theta_j} = \lambda V^{-1}(\theta_j - \hat{\theta}_j), \qquad j = 1, \ldots, p, \tag{3.2.3}$$

$$\tfrac{1}{2}(\theta - \hat{\theta})' \, V^{-1}(\theta - \hat{\theta}) = L_c, \tag{3.2.4}$$

and solve for λ by substituting for $\theta_j - \hat{\theta}_j$ in (3.2.4) to obtain

$$2L_c\lambda^2 = \left[\frac{\partial g}{\partial \theta}\right]' V \left[\frac{\partial g}{\partial \theta}\right]. \tag{3.2.5}$$

If g is linear in θ, $\partial g/\partial \theta_j$ is constant, and the solution is at

$$\theta^* + \frac{1}{\lambda} V \frac{\partial g}{\partial \theta_j}, \qquad (3.2.6)$$

where g can be evaluated. If g is nonlinear the solution is iterative because $\partial g/\partial \theta$ will change with θ. If $L(\theta)$ is not quadratic then θ^* will not lie on the critical contour. In a practical algorithm (see Section 5.5.2) therefore there are a range of options:

(i) to accept the noniterative solution in the interests of speed;
(ii) to iterate for nonlinear $\partial g/\partial \theta$ but to assume $L(\theta)$ is quadratic;
(iii) to adjust each iteration to ensure that θ^* is on the critical contour; or
(iv) to adjust for changes in $\partial L/\partial \theta$ if the parameters are very nonlinear.

The algorithm may fail:

(a) if $g(\theta)$ is discontinuous, for example, $g(\theta) = \theta_1/\theta_2$ when θ_2 can be negative or positive;
(b) if $g(\theta)$ has an optimum within the interior of the critical region, for example, if $g(\theta) = \sum(\theta_j - \hat\theta_j)^2$; or
(c) if the critical contour is discontinuous or concave.

In such cases confidence sets may be discontinuous or unbounded.

EXAMPLE 3.11 (Ross (1978)). The three-parameter exponential

$$E(y) = \alpha + \beta \exp(-kx) = f(x)$$

was fitted to the following data:

x	1	2	3	4	6	8
y	3.2	7.9	11.1	14.5	16.7	18.3

and exact limits of each parameter were found on the critical contour RSS = 5.417, with the following results:

	Minimum	Estimate	Maximum
α	16.89	19.77	26.71
β	-28.61	-23.63	-19.45
k	0.164	0.349	0.572

An approximate solution was obtained using the stable ordinate parameter system

$$\phi_1 = f(1), \qquad \phi_2 = f(4), \qquad \phi_3 = f(7),$$

from which the Lagrange technique was applied to give the following intervals:

	Minimum	Maximum
α	16.90	26.66
β	−28.61	−19.51
k	0.165	0.572

The discrepancies between the two methods are due to the slight departure from the quadratic form of the $L(\phi)$.

In general, the approximate algorithm becomes more reliable as sample size increases, and the relative size of the critical region diminishes. An occasional tendency to oscillate may be counteracted by use of Aitken's d^2 method of accelerated convergence.

3.2.2. Confidence Intervals for Implicit Functions

When $g(\theta)$ is difficult or tedious to evaluate directly it may still be possible to estimate limits by noniterative approximation. The advantage is that, in going from unstable parameters to stable parameters, we may wish to improve the quadratic approximation to $L(\theta)$ using easily computed functions without having to evaluate $g(\theta)$ very frequently. This is achieved by inverting the square Jacobian matrix

$$J = \left[\frac{\partial \theta_i}{\partial g_j}\right],$$

because J^{-1} is the same as

$$\left[\frac{\partial g_j}{\partial \theta_i}\right].$$

For example, in curve fitting it is not difficult to find empirically a set of ordinates at which the expected values are pairwise uncorrelated, or nearly so. Given formulas for the expectations ϕ_i and their derivatives with respect to θ an optimization procedure can be used to find sets for which the sum of squared covariances is minimized, and if this minimum is close to zero then the position of the ordinates will be optimal, although the solution is not always unique. The matrix J^{-1} is then derived and limits on the original parameters θ as functions of ϕ_i are obtained. It remains only to solve the equations

$$\phi(\theta) = \phi_c,$$

where ϕ_c are the sets corresponding to the limiting cases. These equations may be solved iteratively from the known central values, θ, using the matrix J iteratively in the equation

$$\theta^{(r+1)} = \theta^{(r)} + J^{-1}(\phi_c - \phi(\theta^{(r)})).$$

These equations will of course not converge if the confidence contour is not finite, as in the cases described in Section 3.2.1. The algorithm is described in Section 5.5.3.

EXAMPLE 3.12 (Stable Ordinates for Exponential Curves). In the previous example (3.11) the model

$$E(y) = \alpha + \beta \exp(-kx)$$

was fitted to six points. An optimization algorithm converged to the following positions of stable ordinates:

No.	Position of x	Ordinates, ϕ	S.E.	correlations		
1	1.329	4.912	0.296	1		
2	4.274	14.453	0.239	0.0013	1	
3	9.069	18.774	0.431	0.0006	0.0003	1

Using these values critical sets of θ values were estimated noniteratively using equations (3.2.5) and (3.2.6). The corresponding values of the defining parameters were then estimated as follows:

	Minimum	Maximum
α	16.69	25.02
β	−28.21	19.66
k	0.185	0.640

These results are inferior to the solutions obtained in Example 3.11, particularly the estimate for k. However, the relationships between the parameter sets are such that a small error in ϕ can make a large difference to θ, and it is clear that the solution would be improved by iterating to take into account the changes in $\partial\phi/\partial\theta$. These methods are important because they help to justify the use of stable parameters which themselves are of no particular interest. Routine methods for producing confidence intervals without too much extra computation are of great benefit in showing the inadequacy of the linear normal approximations implicit in a summary that relies only on the dispersion matrix of estimates.

3.2.3. Comparison with Fieller's Theorem

A well-known example of a function of parameters is the LD_{50} in probit analysis, the value of x in a quantal response model at which the estimated response is 50%. Fiducial limits are obtained from Fieller's theorem for the ratio $\mu = \alpha_1/\alpha_2$ of two normally distributed random variables (Finney, 1964).

Let α_1 and α_2 be normally distributed with variances V_{11} and V_{22} and covariance V_{12}. Then the linear form, $\gamma = \alpha_1 - \mu\alpha_2$ where $\hat{\gamma} = 0$, is normally distributed with variance

$$\sigma^2(\gamma) = V_{11} - 2\mu V_{12} + \mu^2 V_{22}, \qquad (3.2.7)$$

so that using the t-distribution, the upper and lower limits of γ are $\pm t\sigma(\gamma)$ and μ must satisfy the equation $\gamma^2 = t^2\sigma^2(\gamma)$ or

$$\mu^2(\alpha_2^2 - t^2 V_{22}) - 2\mu(\alpha_1\alpha_2 - t^2 V_{12}) + \alpha_1^2 - t^2 V_{11} = 0. \qquad (3.2.8)$$

The roots of this equation are then the fiducial limits for μ. For some values of α_1 and α_2 there are no real roots, and all values of μ are included within the limits. If the limits for α_2 include zero, then the limits may be *exclusive* rather than *inclusive*, i.e., μ must be either less than μ_1 or greater than μ_2.

Likelihood-based confidence limits for α_1/α_2 are very similar, since t^2 has the F distribution on one degree of freedom. Equation (3.2.8) is equivalent to finding the maximum and minimum values of α_1/α_2 on the ellipse

$$V_{11}(\alpha_1 - \hat{\alpha}_2)^2 - 2V_{12}(\alpha_1 - \hat{\alpha}_2)(\alpha_2 - \hat{\alpha}_2) + V_{22}(\alpha_2 - \hat{\alpha}_2)^2$$
$$= (V_{11}V_{22} - V_{12}^2)\, t^2, \qquad (3.2.9)$$

as described in Section 3.2.1.

The crucial departure from the exact limits occurs when the log-likelihood is not perfectly quadratic. This is the case when quantal responses are observed and errors are binomially distributed. The slight asymmetry of the log-likelihood relative to its quadratic approximation can make a considerable difference to the confidence limits, as illustrated in Fig. 3.10. The limits correspond to slopes of tangents from the origin to the critical contour, and when that passes close to the origin the differences between the two sets of limits is quite marked. The data for Fig. 3.10 are as follows:

x	n	r
-2	5	0
-1	5	3
0	5	2
1	5	3
2	5	5

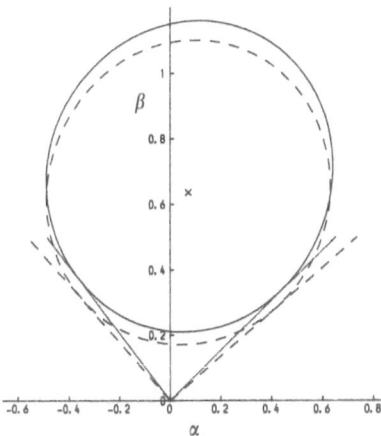

Fig. 3.10. Comparison of likelihood contour (continuous line) with approximation based on asymptotic dispersion matrix (broken line) for quantal response model. The slopes of the tangents from the origin correspond to 95% confidence limits for the LD_{50} based on the likelihood ratio text (continuous line) and Fieller's theorem (broken line).

and the MLEs of α and β in the model $E(r/n) = \Phi(\alpha + \beta x)$ are

$$\begin{bmatrix} \hat{\alpha} \\ \hat{\beta} \end{bmatrix} = \begin{bmatrix} 0.0724 \\ 0.6356 \end{bmatrix}, \qquad \sigma \begin{bmatrix} \hat{\alpha} \\ \hat{\beta} \end{bmatrix} = \begin{bmatrix} 0.2864 \\ 0.2376 \end{bmatrix}, \qquad r(\hat{\alpha}, \hat{\beta}) = 0.502,$$

and the $LD_{50} = -\hat{\alpha}/\hat{\beta} = -0.1139$.

The likelihood-based confidence limits corresponding to the slopes of tangents to the contour $L(\alpha, \beta) = 4.543$ are $(-1.306, +0.971)$ while the corresponding Fieller's theorem limits are $(-1.475, +1.123)$.

3.3. Effective Replication, Influential Observations, and Design

The expression of the log-likelihood function as a sum of squares of deviance residuals (Section 2.3) makes it possible to investigate the following questions:

(a) To what extent does the fitted model give more information than does a single observation? (*Effective replication*)
(b) Which observations give most information on a given parameter or function of parameters? (*Influence problems*)
(c) Which additional observations are needed to improve the precision of estimates of parameters or functions?
(d) Which arrangement of controllable variables is best, either for estimating the parameters in a model or for discriminating between alternative models? (*Nonlinear design problems*)

3.3.1. Linear Regression Models

For a linear regression model

$$E(y) = A\theta, \qquad \sigma^2(y) = V,$$

the log-likelihood is

$$\tfrac{1}{2}(y - A\theta)' \, V^{-1}(y - A\theta), \tag{3.3.1}$$

and the $n \times p$ matrix A is known as the *design matrix*. The $p \times p$ matrix, $C = A'V^{-1}A$, is the *information matrix* and its inverse C^{-1} is the *dispersion matrix* of the estimated parameters. C is assumed to be of full rank.

Since $\hat\theta = C^{-1}A'V^{-1}y = By$, the $p \times n$ matrix B is a dual matrix to the design matrix, its element b_{ij} being the change in θ_i due to unit change in y_j. It has no agreed name, but occurs in the study of influence functions (Belsley *et al.*, 1980). The fitted values are

$$\hat y = A\hat\theta = AC^{-1}AV^{-1}y = Hy, \tag{3.3.2}$$

where the $n \times n$ matrix H has been termed the *hat matrix* (because it puts the hat on y) which has the following properties:

(i) H is idempotent, because $HH = H$.
(ii) $\mathrm{Tr}(H) = p$ since $\mathrm{Tr}(AC^{-1}A^1V^{-1}) = \mathrm{Tr}(C^{-1}A'V^{-1}A) = \mathrm{Tr}(I_p)$.
(iii) The dispersion matrix of y is $HV = AC^{-1}A$.
(iv) Elements of H cannot exceed 1. When $p = n$, H becomes the unit matrix I_n and predicted values equal observed values whatever the model.
(v) A linear transformation of θ does not affect H, because y is unchanged.

Diagonal elements of HV are the variances of fitted values and since $\mathrm{Tr}(H) = p$ their mean value when $V = I_n\sigma^2$ is $p\sigma^2/n$. Since the variance of a single observation is σ^2, and the variance of the predicted value $\hat y_i$ is $h_{ii}\sigma^2$, the quantity h_{ii}^{-1} has a direct numerical interpretation: the number of replicated observations that would give the same variance as that derived from the model. h_{ii}^{-1} is therefore termed the *effective replication* for the prediction $\hat y_i$ (Ross, 1982, 1987). It is not usually an integer but may loosely be interpreted as the average number of neighboring observations that provide information. If the effective replication is close to 1 the observation is *isolated* or *self-estimating* because $\hat y_i$ depends almost entirely on y_i.

Let V be diagonal. Then if A_j is the jth column of A

$$H = \sum_{j=1}^{p} A_j A_j' C^{jj} V^{jj} = \sum H_j, \tag{3.3.3}$$

which is a sum of p independent components each with unit trace. Each component represents the distribution of information from one degree of freedom associated with fitting the pth parameter. The matrix $(I - H)$ represents the information in the residuals, associated with $n - p$ degrees of freedom. In particular, the variance of y_i is a sum of p nonnegative components due to each parameter.

The structure of H is easily demonstrated for polynomial models using tables of orthogonal polynomials (e.g., Fisher and Yates, 1963), where, assuming $V = I\sigma^2$,

$$h_{ij} = \sum_{k=0}^{p-1} \frac{P_k(x_i) \, P_k(x_j)}{\sum_i P_x^2(x_i)}. \tag{3.3.4}$$

For example, for five equally spaced points and three fitted terms representing a quadratic polynomial we obtain the following H matrix:

P_0	P_1	P_2	$70H$					
1	-2	2	$14 + 28 + 20$	...		...	...	...
1	-1	-1	$14 + 14 + 10$	$14 + 7 + 5$	...	...	...	...
1	0	-2	$14 + 0 - 20$	$14 + 0 + 10$	$14 + 0 + 20$	...	...	...
1	1	-1	$14 - 14 - 10$	$14 - 7 + 5$	...	...	...	...
1	2	2	$14 - 28 + 20$	...		...	...	...

$$H = \begin{pmatrix} 0.886 & 0.257 & -0.086 & -0.143 & 0.086 \\ 0.257 & 0.371 & 0.343 & 0.171 & -0.143 \\ -0.086 & 0.343 & 0.486 & 0.343 & -0.086 \\ -0.143 & 0.171 & 0.343 & 0.371 & 0.257 \\ 0.086 & -0.143 & -0.086 & 0.257 & 0.886 \end{pmatrix}$$

The values of h_{ij}^{-1} are (1.13, 2.70, 2.06, 2.70, 1.13) which shows that the outer points y_1 and y_5 are nearly self-estimating.

3.3.2. General Models

To extend the foregoing results to nonlinear models and to nonnormal error distributions consider the log-likelihood as a sum of squares of deviance residuals

$$L(\theta) = \tfrac{1}{2}\sum e_i^2(\theta) = \tfrac{1}{2} e'e. \tag{3.3.5}$$

Expanding in Taylor series about the MLE θ,

$$L(\theta) = L(\hat\theta) - e' A \delta\theta + \tfrac{1}{2} \delta\theta'(A'A + D)\,\delta\theta + O(\delta\theta), \tag{3.3.6}$$

where $\delta\theta = \theta - \hat\theta$

$$A = -\left(\frac{\partial e}{\partial \theta}\right)_{\theta=\theta_j} \quad \text{and} \quad D = \frac{\partial^2 e_i}{\partial \theta_j \partial \theta_k} e_i \quad \text{and} \quad e' A\hat\theta = 0.$$

The $n \times p$ matrix A is given a negative sign so that for a linear regression model it corresponds with the design matrix A, multiplied if necessary by the factor V^{-1}.

The matrix D is often ignored on the grounds that it is usually small relative to $A'A$, but examples can be found where its effect is large, either positive or negative. When the parameters are correlated the effect on the dispersion matrix is largest.

EXAMPLE 3.13. In Example 3.11 (Section 3.2.1) we find that the dispersion matrix of (α, β, k) including D is

$$\begin{pmatrix} 0.5172 & & \\ -0.0369 & 0.6650 & \\ -0.024429 & -0.00935 & 0.001398 \end{pmatrix}$$

and excluding D is

$$\begin{pmatrix} 0.5295 & & \\ -0.0324 & 0.6668 & \\ -0.02499 & -0.00961 & 0.001439 \end{pmatrix}.$$

The advantage of ignoring D in the following discussion is that the matrices B and H may be more easily interpreted. The matrix V becomes the identity matrix I_n because the distribution of y is taken into account in the definition of deviance residuals.

The matrix $A = -(\partial e/\partial \theta)_{\theta=\theta_j}$ is not constant, as in the linear case, but parameter-dependent and therefore only computable *a posteriori*. It may be called the *effective design matrix*, to emphasize the analogy without implying that it is constant. The matrix

$$B = (A'A)^{-1}A' = C^{-1}A' \tag{3.3.7}$$

expresses adjustments to the parameter estimates in terms of residuals, since at convergence

$$\delta\theta = Be.$$

Rows of B may be called *parameter loadings*, expressing the effect of changes in each observation (through its deviance residual) on the estimated parameters.

EXAMPLE 3.14. For the exponential curve fitted in Example 3.11 the matrices A and B' are as follows:

		A			B'	
x	$\hat{\alpha}$	$\hat{\beta}$	$\hat{k}$	$\hat{\alpha}$	$\hat{\beta}$	$\hat{k}$
1	1	0.705	16.67	0.215	0.663	−0.0185
2	1	0.498	23.52	−0.177	0.175	0.0100
3	1	0.351	24.88	−0.248	−0.090	0.0182
4	1	0.248	23.40	−0.151	−0.220	0.0156
6	1	0.123	17.47	0.212	−0.282	−0.0018
8	1	0.061	11.50	0.568	−0.246	−0.0207

The columns of matrix B' show that $\hat{\alpha}$ is strongly influenced by the values of y at $x = 8$, whereas $\hat{\beta}$ depends more on the value of y at $x = 1$. $\hat{k}$ is seen to be a measure of curvature that depends on the contrast between the inner and outer points on the curve.

The hat matrix H is then found to be

$$
\begin{pmatrix}
0.882 & 0.259 & -0.031 & -0.130 & -0.060 & 0.087 \\
0.259 & 0.325 & 0.294 & 0.217 & 0.031 & -0.126 \\
-0.031 & 0.294 & 0.380 & 0.341 & 0.122 & -0.106 \\
-0.130 & 0.217 & 0.341 & 0.343 & 0.201 & 0.028 \\
-0.069 & 0.031 & 0.122 & 0.201 & 0.319 & 0.395 \\
0.087 & -0.126 & -0.106 & 0.028 & 0.395 & 0.722
\end{pmatrix}
$$

and h_{ii}^{-1} (1.13, 3.08, 2.63, 2.92, 3.13, 1.39). Note that the first point ($x = 1$) is more isolated ($h_{ii}^{-1} = 1.13$) than the last point ($x = 8$, $h_{ii}^{-1} = 1.39$), while inner estimates of $\hat{y}$ depend on roughly three neighboring values of y.

3.3.3. Designs for Nonlinear Estimation

The term "design" in this context is the choice of values of x which affect the precision of the estimates of parameters and functions, and the ability of the data to discriminate between models. The choice of designs is usually constrained by the sample size and the practicable range of x values.

From the analysis of a given data set it is possible to investigate different distributions of the x values within the same range, assuming the model to be true and the estimated parameters to be the true parameters. This is the approach of the classic work on optimal design (Box and Lucas, 1959; Box, 1968; Fedorov, 1972). The general procedure is to define a criterion such as

$$D(x) = \text{Det}(V(\hat{\theta})|x),$$

which is the determinant of the dispersion matrix of $\hat{\theta}$, and to find the D-optimal design, x_D, that minimizes $D(x)$. The advantage of D-optimality is that it can be shown to be independent of the parametrization, because if we redefine

$$\phi = T\theta$$

for some linear transformation expressed by the matrix T, then

$$V(\phi) = T'V(\theta)T,$$

and

$$\text{Det } V(\phi) = \text{Det}(T)^2 \cdot D(x),$$

where $\text{Det}(T)$ is independent of x. To a first-order approximation the same is true of nonlinear transformations.

In practice, it is found that for functions of a single variable x the D-optimal design invariably turns out to be a set of p different values of x replicated as equally as possible. If n/p is not an integer the number of observations at each point differs by at most 1. The expectations at the chosen points may be considered as a set of transformed parameters, and the hat matrix, H, is diagonal with values of h_{ii}^{-1} which are the integers closest to n/p. It is easy to show that if the design consists of p groups of points the determinant $D(x)$ is minimized when the replication is as even as possible, and that it is irrelevant which values of x receive which replication.

The reason D-optimal designs are not used to any great extent is that they depend crucially on prior knowledge of the likely values of the parameters, and are therefore only appropriate for the final stages of estimation in a situation which is readily repeatable, such as the determination of physical constants. In other contexts we would prefer designs which are reasonably efficient for a wide range of parameter estimates and which allow models to be compared.

An informal procedure would be to examine the hat matrix H and to modify the design so as to decrease the variability of h_{ii}^{-1}. But since the rows of H correspond to points in the new design rather than the original design, it may be preferable to consider the variances of predicted values of y at a set of fixed values of x, otherwise the procedure simply leads back to the D-optimal design.

A better procedure is to consider the benefits of observing additional points (assuming the cost of observation is not too great) and to reinforce the design where h_{ii}^{-1} is low. For example, 3.14, this would mean that extra observations would be most useful in the neighborhood of $x = 1$ and $x = 8$, or beyond.

If two models are under consideration the design should be reinforced at points where the predicted values differ most. In many cases this implies extrapolation beyond the given range of data. In the examples in Sections 6.2.1 and 6.2.2 there is no possibility of discriminating between exponential and logistic growth over a range of years in which growth is essentially exponential. The required extra design points are unfortunately not observable until after the time at which the estimates are needed.

EXERCISES

1. The "two-hit" model, $E(y) = (1 + \theta x) \exp(-\theta x)$, is fitted to two points, $(1, y_1)$ and $(2, y_2)$. Assuming independent normal errors plot the solution locus and common estimate classes for θ in the space of y_1 and y_2.
2. Evaluate the matrices B and H for the models
    (i) $E(y) = \alpha + \beta x + \gamma x^2$,
    (ii) $E(y) = \alpha + \dfrac{\beta}{1 + \gamma x}$,

    fitted to the data of Example 3.1, and compare the results with those for the exponential model.

CHAPTER 4

The Geometry of Nonlinear Inference

4.1. The Role of Graphical Representations

Graphics have been used in earlier chapters to illustrate various aspects of nonlinear modeling. In this chapter graphical methods will be discussed systematically, to show how graphs may be used to help select a model, to diagnose problems, and to understand the relationships between data and parameters.

If we assume that there are the observations $y_1, \ldots, y_n$ with associated variables $x_i, \ldots, x_n$, and a model with p parameters θ, then there are three distinct classes of graphical representation:

(a) $(x-y)$ plots, in which observations y (or functions of observations or expectations) are displayed in relation to one or more of the x variables. These plots are usually in two or three dimensions, depending on the number of x variables involved.

(b) $(x-y)$ plots, in which vectors of observations y or predictions $\hat{y}$ are plotted against each other in n-dimensional space. Theoretical discussions illustrate two or three dimensions and invite the reader to generalize to n dimensions. Practical use of such plots is only possible for very small samples, or for typical pairs of observations or functions of observations.

(c) $(\theta-\theta)$ plots, in which log-likelihoods or other functions are plotted in p-dimensional space. Up to four parameters at a time may reasonably be represented as a two-way array of contour plots.

Graphs appeal to visual intuition, by which certain shapes and patterns are recognized and others rejected. We can learn to recognize the general characteristics of shapes: linearity, positive or negative curvature, asymptotes, maxima and inflexions, ovalness, and continuity. But we cannot easily distinguish an exponential curve from a rectangular hyperbola, an ellipse from another type of oval, or a Gaussian curve from other bell-like shapes. The general principles of graphical representation are:

(1) transformation of scales to provide recognizable shapes; and
(2) subtraction of expected shapes from observed shapes to reveal differences.

4.2. Data Plots in $(x-y)$ Space

Plotting the data is the starting point of model selection. The scatter diagram of y against a single x variable suggests not only the form of the curve but also the distribution of residuals, the range of the data, the pattern of the design, and any other particular features such as clusters or outliers that might require explanation (see e.g., Fig. 2.4).

If there are several x variables with regularly spaced values the data may be displayed by plotting y against $x_1 + c_2 x_2 + c_3 x_3, \ldots$, etc., where c_2 and c_3 may be chosen to separate the variables as required. In Fig. 4.1 two variables are shown so that individual responses to x_1 and x_2 may be inferred separately. The suggested model is Baule's equation, the product of two exponential functions (Patterson, 1959). The same data may be displayed on a grid with empirical contours drawn (Fig. 4.2) but provides less insight into possible model forms; recognition of contour shapes is not easily learnt, especially for nonlinear models. The main messages from contour diagrams are the directions of trends, the type of interaction between the variables, and the degree of complexity required of the model.

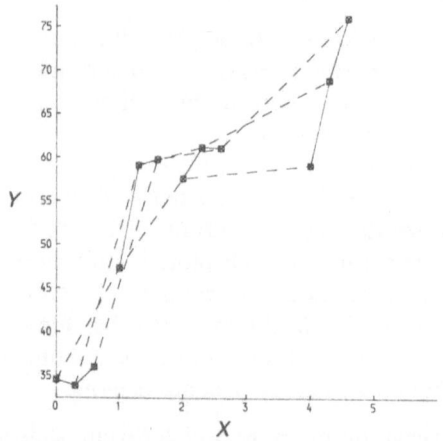

Fig. 4.1. Simultaneous plot of yield of grass (Y) against fertilizer combination $X = P + 0.34K$ (data from Patterson, 1969). Responses to P shown as broken lines and to K as continuous lines.

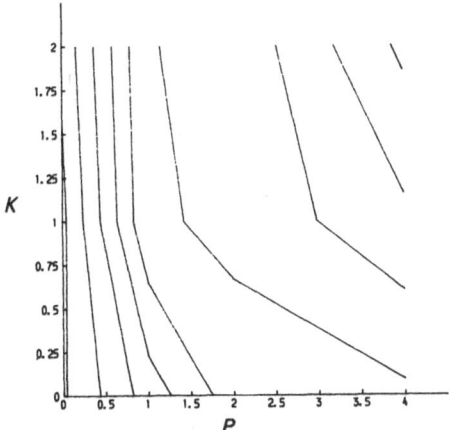

Fig. 4.2. Empirical contours for the data in Fig. 4.1 using linear interpolation between data values. Contour levels are 35(5)75.

4.2.1. Plots on Transformed Scale of y

Transformations of the y scale to simplify the shape of response are well known. The function $y = f_1(f_2(x))$ is transformed to

$$f_1^{-1}(y) = f_2(x),$$

where f_2 is a simple function such as a straight line, and f_1^{-1} may be $1/y$, $\log(y)$, or $\log(y/(1-y))$, for example. Special graph paper may be used to plot x against functions of y, and in standard applications use of log or probability paper used to be widespread before the computer age. A printer's catalogue lists various combinations of log, reciprocal, logistic, probability, and Poisson scales while computer graphics packages can produce whatever transformations are appropriate.

The logistic curve,

$$y = a + \frac{c}{1 + \exp(-b(x - m))}, \tag{4.2.1}$$

may be rewritten

$$\log\left[\frac{y - a}{a + c - y}\right] = b(x - m), \tag{4.2.2}$$

which suggests that, given plausible estimates of a and c, it is possible to estimate b and m graphically by plotting $\log(y^*/(1 - y^*))$ against x where

$$y^* = \frac{y - a}{c}. \tag{4.2.3}$$

In Fig. 4.3, the data of Table 4.1 are plotted with two different choices of a and c. This illustrates two problems:

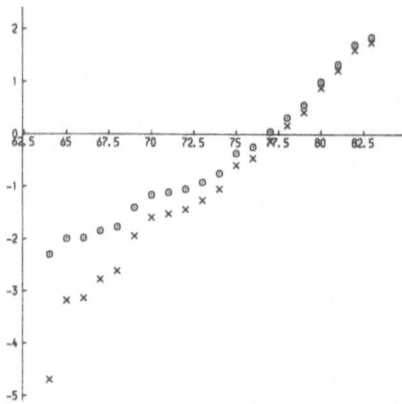

Fig. 4.3. Plot of $\log((y - y_{min})/(y_{max} - y))$ against x, where $y =$ United Kingdom passengers by sea (millions), $x =$ year $- 1900$. (a) $y_{min} = 3$, $y_{max} = 15$; (b) $y_{min} = 4$, $y_{max} = 15$ (data of Table 4.1).

(i) Points close to asymptotes are greatly affected by transformations to improve linearity: points beyond the assumed asymptotes may not be transformed at all.

(ii) The distributions of residuals, and hence judgments about departures from linearity, are similarly affected.

The graphs therefore serve only to check on the plausibility of the model and to give rough initial estimates of parameters. Direct least squares estimation using the transformed scale is commonly performed but is liable to be misleading. Use of the reciprocal scale is particularly common, as for example,

$$y^{-1} = \text{polynomial in } x,$$

$$y^{-1} = \text{polynomial in } (x^{-1}),$$

$$y^{-1} = a + b \exp(-kx).$$

Table 4.1. U.K. passengers by sea (millions), 1964–83.

Year	Number of passengers	Year	Number of passengers
1964	4.10	1974	6.87
1965	4.44	1975	7.94
1966	4.46	1976	8.28
1967	4.65	1977	9.15
1968	4.76	1978	9.97
1969	5.39	1979	10.66
1970	5.88	1980	11.80
1971	5.99	1981	12.51
1972	6.13	1982	13.17
1973	6.44	1983	13.38

SOURCE: *Annual Abstracts of Statistics*, HMSO.

The last example is once more the logistic curve, and an example is discussed by Brooks *et al.* (1978). In their example the ratio of cars per head of population is assumed to grow with time according to a logistic curve. The curve is fitted directly and inversely, with very different parameter estimates because the variances of the residuals vary with the fourth power of y, causing very large changes in weighting. A better treatment is to assume that log y is equally weighted, which gives the same result using y^{-1} because log y^{-1} is $-$log y. This example is discussed more fully in Section 6.2.2.

4.2.2. Plots on Transformed Scale of x

Transformations of the x scale are not often employed, yet they can be more illuminating than transformations of the y scale. Their advantage is that the y coordinate, and hence the visual distribution of residuals, is unaffected by the transformation.

For curves of the form

$$y = a + bf(x, \theta)$$

suitable trial values of θ are chosen, and $x^* = f(x, \theta)$ is used as an x axis. For comparison with other values of a it may be desirable to standardize the range of x^*, plotting the highest and lowest values at the same place on the graph. This avoids difficulties when the parameters change rapidly. For example, if

$$f(x, \theta) = \exp(-kx), \qquad x = 1, 2, 3, 4,$$

the following table shows how x^* can be standardized by the formula

$$x^* = \frac{f(x) - f(x_1)}{f(x_4) - f(x_1)}.$$

$x=$	1	2	3	4	1	2	3	4
k	$f(x)$				x^*			
0.1	0.9048	0.8187	0.7408	0.6703	0	0.367	0.699	1
0.5	0.6065	0.3679	0.2231	0.1352	0	0.506	0.814	1
1.0	0.3678	0.1353	0.0498	0.0183	0	0.665	0.910	1

The y values, once plotted, are easily replotted for different values of θ, giving a quick check on the linearity of the transformed response.

The logistic curve requires two nonlinear parameters k to be guessed: the function

$$f(x, b, m) = \frac{1}{1 + \exp(-b(x - m))}$$

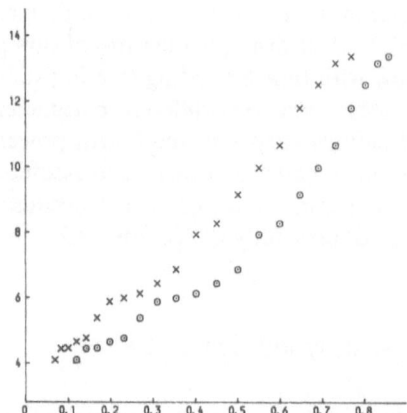

Fig. 4.4. Plot of y against $1/(1 + \exp(-b(x - m)))$ for the data of Table 4.1. (a) $b = 2$, $m = 77$; (b) $b = 2$, $m = 74$.

lies between 0 and 1, and the curve's asymptotes, a and $a + c$ in the notation of (4.2.1), may be read off the graph as intercepts at $f = 0$ and $f = 1$. The data of Table 4.1 are plotted in this way in Fig. 4.4, showing a reasonably linear response for the trial values chosen. The extreme case as m becomes very large is the exponential, $f(x, b) = \exp(bx)$, which must be standardized for comparison. When m is large and negative, $\exp(-bx)$ should be used. These limiting cases are useful when it is not clear that a logistic curve will fit at all. If there is a fixed lower asymptote, the line must be drawn through the origin.

Routine use of $\log x$ or $1/x$ instead of x is used when the transformation is known in advance, as in bioassay where the x variate is usually plotted on a log scale. However, it is not always recognized that certain curves are equivalent under transformation of the x scale, such as

$$y = a + b \exp(-kx),$$

and

$$y = a + bx^c,$$

because the latter can be written $y = a + b \exp(c \log x)$. Similarly, the curves

$$y = \frac{ax + bx^2}{1 + cx + dx^2},$$

and

$$y = \frac{a + bx}{1 + cx + dx^2},$$

are equivalent after the transformation $x^* = 1/x$ (with different values of a, b, c, and d, of course).

4.2.3. Difference and Ratio Plots

Some models give equations in which expectations for equally spaced data have particular relationships. Differences or ratios of successive observations can be plotted in such a way that parameters may be estimated and departures from the model recognized.

Discrete distributions recording the frequencies n_r of r events, where $r = 0$, $1, 2, \ldots$, may be displayed to indicate whether or not the Poisson distribution is likely to hold. Since for the Poisson distribution with parameter m,

$$p_r = \frac{e^{-m} m^{-r}}{r!},$$

$$p_{r+1} = \frac{m p_r}{r+1}.$$

Hence, if we plot $m(r) = (r + 1)n_{r+1}$ against r, we should obtain a series of estimates of m, as suggested by Gart (1970). For Poisson data the scatter of values should be horizontal about the line $m(r) = m$. The negative binomial distribution with parameter k satisfies the recurrence relationship

$$p_{r+1} = \frac{m}{m+k} \frac{r+k}{r+1} p_r,$$

where $m(r)$ is an estimate of $m(1 + (r - m)/(m + k))$, and the trend of points should rise. An alternative, simpler method is to plot $\log(n_r)$ and $\log((r + 1)n_{r+1})$ simultaneously. The two curves remain a constant distance apart if the data fits a Poisson distribution; small values of n_r should be ignored. Two examples are given in Fig. 4.5, the famous data of Rutherford and Geiger (1910) of counts of alpha particles, and Pearson's data on defective teeth in boys (Pearson and Moul, 1925).

For empirical curve fitting, equally spaced points must be selected or estimated. The curve

$$y = a + bf(x)$$

can be studied in terms of the functions $y_i - y_j$ eliminating a, and $(y_i - y_j)/(y_j - y_k)$ eliminating b. For the exponential curve, $(y_{i+2} - y_{i+1})/(y_{i+1} - y_i)$ should be constant. For the logistic curve, use $1/y$, and for the Gompertz curve use $\log y$ in the above expressions. Points should not be so close together that the ratios are too variable: if necessary, several series may be used. If the plotted points are not on a horizontal line, alternative forms are indicated, such as the sum of two exponentials.

A plot of y_{i+1} against y_i may also reveal a relationship. For the exponential curve the plotted points lie on a straight line (tending to cut the line $y_i = y_{i+1}$ at $y_i = a$). An example is given in Fig. 4.6. An alternative form of this plot is to draw two vertical lines, $x = x_1$ and $x = x_2$, and to join the points (x_1, y_i) and (x_2, y_{i+1}) for several values of i. These lines when produced should meet on the line $y = a$. This plot is particularly useful for estimating the power parameter in the generalized logistic curve, fitted to four equally spaced points.

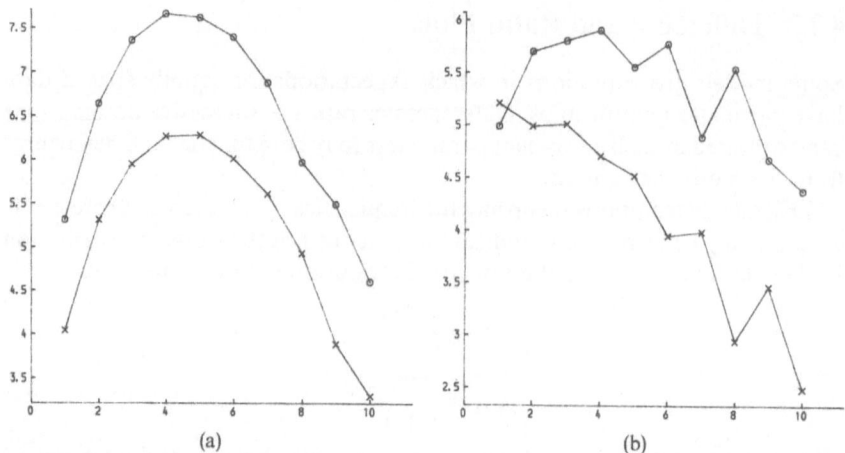

Fig. 4.5. Plots of $\log(n_r)$ and $\log((r + 1) n_{r+1})$ against $r + 1$ as a test for the Poisson distribution (a) Alpha particles (Rutherford and Geiger, 1910); (b) defective teeth in boys (Pearson and Moul, 1925).

For trial values of the power parameter θ, the values of $y^{-1/\theta}$ may be plotted, and it is easy to see whether the three lines in Fig. 4.7 are likely to be concurrent for some value of θ, and if so whether the estimated asymptote is positive and valid.

Other examples of difference and ratio plots are as follows:

(1) If the curve is a quadratic function, then $y_{i+1} - y_i$ against i should be a straight line.

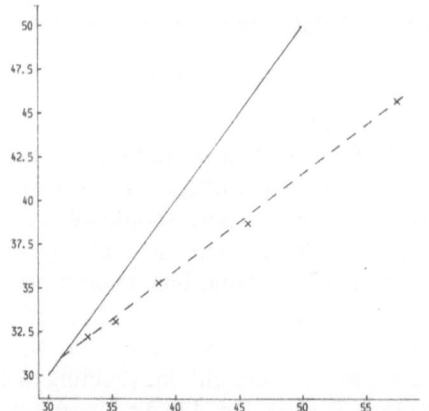

Fig. 4.6. Plot of $y(x + 1)$ against $y(x)$ for equally spaced observations likely to fit an exponential curve. Data from Stevens (1951) represent temperatures (°F) of cooling water at half-minute intervals. The asymptotic temperature is at the intersection of the data trend line with the line $y(x + 1) = y(x)$.

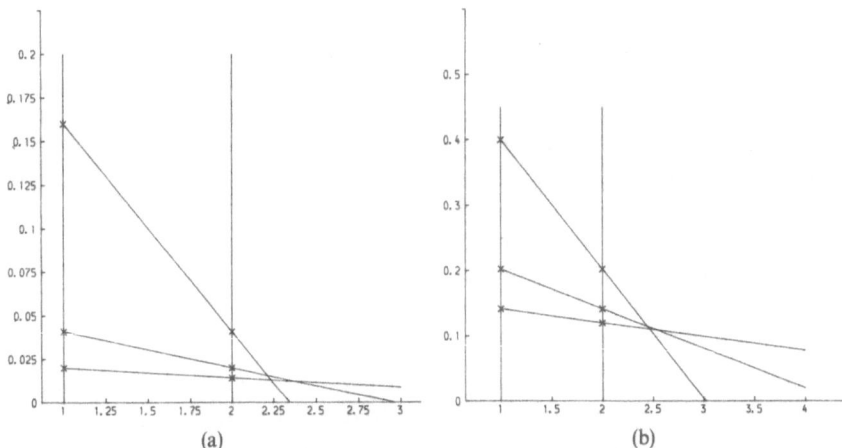

(a)   (b)

Fig. 4.7. Nomograms for estimating parameters of a generalized logistic curve through four equally spaced points. Data adapted from Nelder (1961) represent growth of carrots related to day degrees, with successive values, $y = 6.25, 24.51, 50.0, 69.34$. Lines joining successive values of y^{-t} should coincide if t is the correct estimate. (a) $t = 1$, representing an ordinary logistic curve; (b) $t = 5$, showing a better fit.

(2) If the curve is a rectangular hyperbola, with asymptote $y = a$, then differences $(1/(y_{i+1} - a) - 1/(y_i - a))$ should be constant.

Second differences can also be used if the data are sufficiently accurate. Changes in second differences reveal asymmetry in the curve.

4.2.4. Residual Plots

Much has been written on the subject of residual plots in linear modeling (see, e.g., Barnett and Lewis, 1978; Hawkins, 1980; McCullagh and Nelder, 1982; Atkinson, 1982), and their use is important in nonlinear modeling also.

Having fitted a model, the residuals should be a random sample from the assumed error distribution, taking into account the loss of degrees of freedom attributable to fitting. Residual plots aim to reveal any systematic departures from the assumptions of the model, by displaying possible trends, changes in variance, and the presence of outliers.

In the linear model with normal errors,

$$y = A\theta + N(0, \sigma^2),$$

the dispersion matrix of residuals is

$$(I - A(A'A)^{-1}A')\sigma^2 = (I - H)\sigma^2, \tag{4.2.4}$$

where H is the hat matrix (discussed in Section 3.3) in relation to effective replication. The variance factors are the diagonal elements of $I - H$, which sum to $n - p$, so that the $(n - p)$ degrees of freedom are distributed unequally between the n residuals. The outer residuals of a curvilinear relationship have lower variance than those in the center, especially in small samples, unless the design has been adjusted to stabilize the variance (Section 3.3.3). Barnett and Lewis (1978) suggest standardizing the residuals to equal variance before plotting, but even that procedure does not necessarily show the outer residuals (with lower variance) to be unusual, because a change in the observed value is partly taken up by a change in the fitted value. In large samples outliers may be apparent, but in small samples the standardized residual is not significantly large.

The main types of residual plots are as follows:

(1) Residuals plotted against independent variables (x) to reveal departures from the model, correlations with neighbours, and outliers.
(2) Residuals plotted against observed or predicted values (y) to reveal changes in variances.
(3) Ranked residuals plotted against probability percentage points, to test for normality or other distributional assumptions.

4.2.5. Residuals Plotted Against Independent Variables

Residuals plotted against a single variable, x, readily show any pattern that might suggest an additional term in the model. The nature of the additional term is not so easily predicted, however. In curve fitting a hierarchical series of models may be tried, each with one more parameter. The difference between the fitted $(p + 1)$-parameter curve and the fitted p-parameter curve is usually a function wih p changes of sign. For polynomials it is merely the orthogonal polynomial of degree p. For nonlinear models it is easier to construct the functions empirically than to gain insight from the algebraic formulas of equations such as

$$R(k, x) = e^{-kx} - \alpha(k) - \beta(k)x,$$

which represents the difference between a straight line and an exponential with parameter k. For smooth, continuous curves, the difference function is similar in shape to an orthogonal polynomial.

If the residuals for a fitted p-parameter model exhibit only p changes of sign, they should correlate highly with some suitable difference function. In Fig. 4.8 residuals to a three-parameter hyperbola are compared with the difference function for a four-parameter hyperbola, showing a marked improvement in the fit. Conversely, with more than p changes of sign correlations are likely to be lower, and the fit must be considered adequate.

To represent the pattern of residuals with two independent variables a scatter diagram is required, with points labeled by sign and rounded value of the residual: patterns of positive and negative residuals may suggest the appropriate form of model, but this skill is not easily acquired. In Fig. 4.9 the

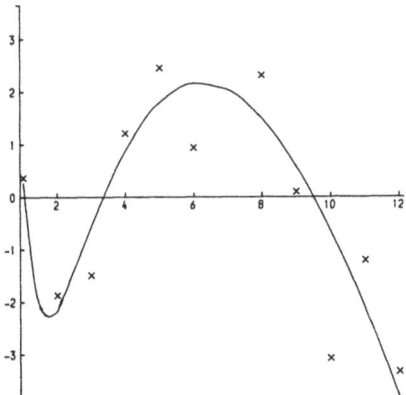

Fig. 4.8. Residuals from a three-parameter hyperbola plotted alongside a curve which is the difference between fitted four-parameter and three-parameter curves. The shape is roughly similar to that of a cubic polynomial.

residuals from a plane fitted to the data of Figs. 4.1 and 4.2 suggest curved response at an oblique angle, indicating the need for a model able to reflect the curvature in the direction $x_1 + x_2 = $ constant.

Serial correlation is observed when consecutive observations come from the same source and change is systematic. The sign may change too frequently for an extra term in the model to be of use, but less frequently than for completely independent observations. If detected, serial correlation can be measured in the normal way, and the model fitted again with allowance for the correlation. This is only practicable with equally spaced values except in very small samples.

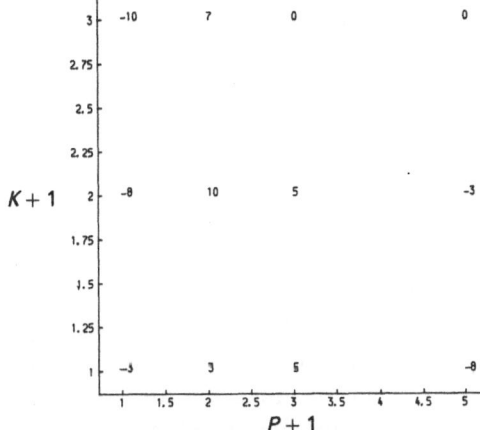

Fig. 4.9. Residuals after fitting a plane to the data of Fig. 4.2, showing positive and negative values which suggest the need for models to account for the curvature of the surface.

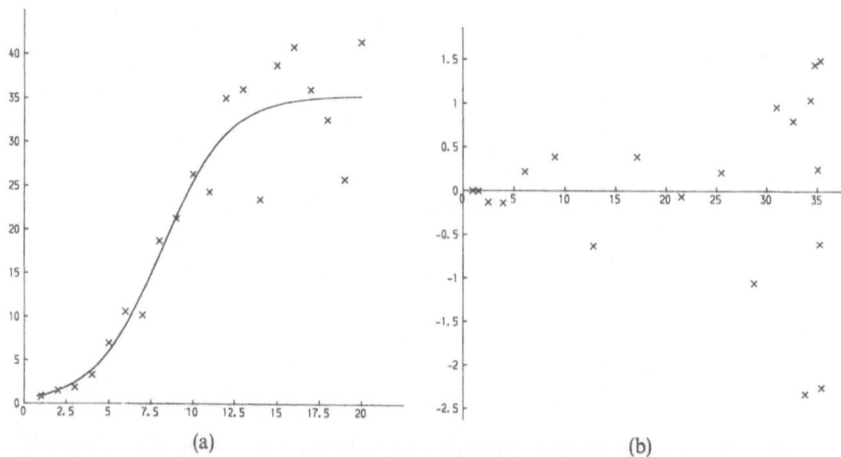

(a)   (b)

Fig. 4.10. Data simulated from a logistic curve with residual variance proportional to expectation. (a) Data and curve fitted assuming equal variances; (b) residuals plotted against expectations, showing roughly proportional relationship between standard error and expectation.

4.2.6. Residuals Plotted Against Fitted Values

Since plots of residuals against y values are intended to show whether the variance assumptions hold, it may be desirable to standardize using formula (4.2.4). Then if some relationship,

$$\text{Var}(y - \hat{y}) = f(\hat{y}),$$

is suggested, a further standardization by $f^{1/2}$ could be employed.

In Fig. 4.10 standardized residuals are shown for a model fitted with equal variances, when it is clear that the assumption of proportionality with $\hat{y}^2$ should be a great improvement. In practice, the model could then be fitted again and the pattern of residuals might change, but if the residuals are small the form of relationship is likely to remain the same.

4.2.7. Probability Plots

The normal plot technique is to rank the residuals and plot them against percentage points of the normal distribution. The rth residual out of n is plotted against the normal equivalent deviate of the function $r/(n + 1)$. If the distribution is normal the points should lie roughly on a straight line through the origin with slope σ. If the distribution is very skew the line will not pass through the origin, and if the distribution is long-tailed the line will be curved. If the variances of the residuals are very different the residuals must be standardized by their standard deviations (using formula (4.2.4)). A normal

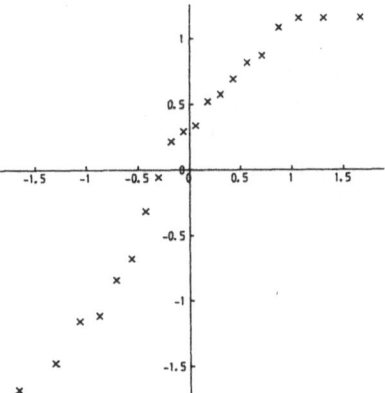

Fig. 4.11. Normal plot of standardized ranked residuals for the data of Fig. 4.10. The x-axis is the normal equivalent deviate of $r/(n+1)$ where r is the rank.

plot of the residuals in Fig. 4.10 is shown in Fig. 4.11, with and without standardization; the standardized plot is of course very close to linearity.

An alternative plot, which can be used for any distribution where percentage points are known, is to plot the cumulative distribution function of the ranked residuals against rank number. This is a transformed version of the previous plot, and has the advantage of finite and fixed size, the unit square. For normally distributed residuals, plot the normal probability integral (using the best estimate of σ^2 available) of the rth residual against $r/(n+1)$: the points should lie close to the 45° line. Deviations from the line can be tested by the Kolmogorov–Smirnov goodness-of-fit test that compares the maximum ab-

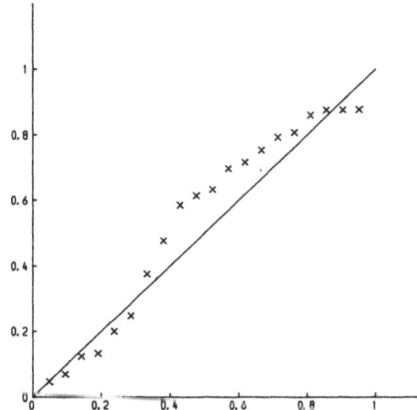

Fig. 4.12. Normal probability integral of the standardized residual plotted against rank, with the 45° degree line to show the extent of any discrepancies. The data are the same as in Fig. 4.11 but the points are equally spaced.

solute deviation with maximum allowable deviation for given sample size. The
data of Fig. 4.11 are plotted in this way in Fig. 4.12.

For models in which the distribution is nonnormal, likelihood-based residuals may be used in a normal plot. Alternatively, the cumulative distribution function for each observed value may be worked out and the results
ranked. For discrete data following the Poisson or binomial distributions it
may be necessary to use the midpoint of each vertical interval on the cumulative distribution plot. These plots require a computer program and are not
the simple type of plot that can be done by hand.

It is not necessary to produce probability plots for every data set of a batch,
but it is useful to check from time to time that the error distribution assumptions are valid for a class of modeling situations.

4.2.8. Plots of Derivatives and Parameter Loadings

The nonlinear regression model

$$E(y) = f(x, \theta)$$

may be further illuminated by plotting simultaneously against x and the
derivative curves

$$\frac{\partial f(x, \hat{\theta})}{\partial \theta_1}, \frac{\partial f(x, \hat{\theta})}{\partial \theta_2}, \dots.$$

Although the scales will differ, the tendency to rise and fall together may be
noted (Fig. 4.13). The correlation between the predicted values of these func-
tions contributes to the correlation between the estimates $\hat{\theta}_1$, $\hat{\theta}_2$, etc., so that

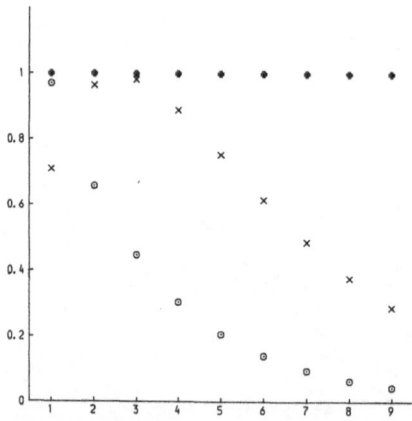

Fig. 4.13. Plots of $df(x, \hat{\theta})/d\theta$ against x for the exponential curve fitted to the data of
Fig. 2.4. The scales are different for each set of points. $(\times)\, x \exp(-\theta_1 x)$; $(\bigcirc) \exp(-\theta_1 x)$;
$(*)\, 1$.

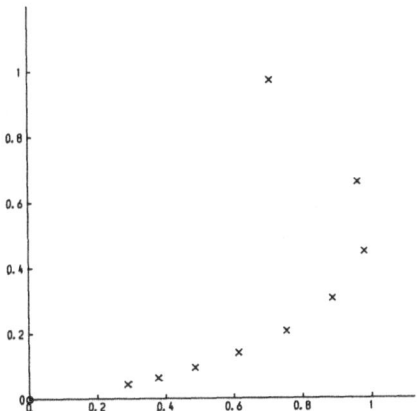

Fig. 4.14. Plot of $\partial f(x_1 \hat{\theta})/\partial \theta_1$ against $\partial f(x_1 \hat{\theta})/\partial \theta_2$ for the example of Fig. 4.13.

in order to reduce correlations by taking further observations, parts of the data range must be used where the functions differ in shape. Since the information matrix is estimated by elements

$$c_{jk} = \sum w_i \frac{\partial f(x_i)}{\partial \theta_j} \frac{\partial f(x_i)}{\partial \theta_k},$$

it is the correlation relative to the origin that is important. An alternative procedure for a particular pair of parameters is to plot $\partial f/\partial \theta_1$ against $\partial f/\partial \theta_2$: highly dependent parameters produce a curve that is almost a straight line through the origin (Fig. 4.14).

Parameter loadings (see Section 3.3.2) are obtained by multiplying the matrix of first derivatives by the dispersion matrix V. The jth parameter loading

$$u_j(x, \theta) = \sum_k v_{jk} \frac{\partial f}{\partial \theta_k}(x, \theta)$$

may be plotted as a function of x to show the parts of the range of x that contribute most to the estimation of θ_j. These plots reinforce the conclusions obtained by studying parameter loadings as *vectors*, that estimation of a parameter requires relevant data which may mean changes in the design if improvements are to be made. Thus, while studies of D-optimal designs (Section 3.3.3) show how points should be allocated within the constraints imposed by the experiment, studies of parameter loadings may show that little further progress can be made without altering the constraints.

In Fig. 4.15 parameter loadings for the two Michaelis–Menten parameters (Section 2.1.4) are shown. Whereas for given data reparametrization is useful to ease estimation, redesign is necessary if the required parameters are to be estimated more independently, for example, by sampling where the value of x is greater than the maximum in the present data.

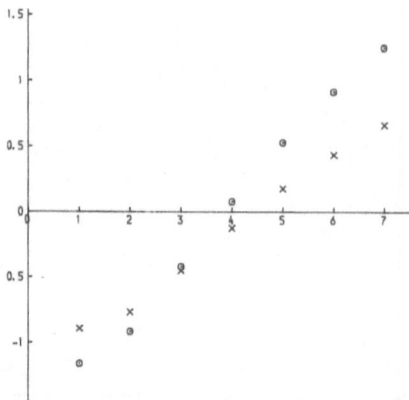

Fig. 4.15. Parameter loadings plotted against x for the parameters of the Michaelis–Menten curve (data of Table 2.1). ($\times$) Loadings for θ_1; ($\odot$) loadings for θ_2. The data points affect both parameters in much the same way, but a larger x should help to provide more information about θ_2 rather than θ_1.

General *function loadings* for functions of parameters $g(\theta)$ are

$$U(x, \theta) = \sum_{j,k} V_{jk} \frac{\partial g}{\partial \theta_j} \frac{\partial f(x, \theta)}{\partial \theta_k},$$

and may also be plotted and interpreted in the same way. Functions are of course merely potential new parameters, and the function loading is not affected by the form in which the other parameters are incorporated into the

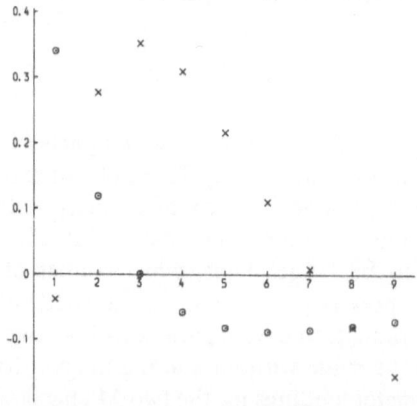

Fig. 4.16. Function loadings plotted against x for the exponential curve (Fig. 2.4). The scales have been adjusted to make the plots comparable. ($\times$) Ordinate at $x = 3$; ($\odot$) slope at $x = 3$.

model. In Fig. 4.16 function loadings are illustrated for the slope and ordinate of an exponential curve at a given point. The slope depends on the contrast between data values at points on either side, while the ordinate depends on values of neighboring data points.

4.3. Data Plots in $(y-y)$ Space

In discussions of model fitting much use is made of the geometrical analogue of regression developed by Fisher (1956), where the data vector y is treated as a point in n-dimensional space. Box (1957) used the term *solution locus* to denote the p-dimensional subspace of exact fits to the model, $f(\theta)$. Least squares estimation is then a matter of finding the point on the solution locus closest to the data point y. Beale (1960) used this representation to discuss measures of nonlinearity in terms of curvature of the local solution locus, and Bates and Watts (1980) have made further contributions to this subject. The term "expectation surface" instead of "solution locus" has been used by Watts (1982).

There is a difference, however, between formal illustrations, in which points and curves in two-dimensional space are said to be points and manifolds in n-dimensional space, and practicable graphs from which detailed conclusions may be drawn. The difficulty is that interesting data sets rarely involve as few as two or three observations, and that lines and curves in two dimensions do not display the full complexity of geometrical reationships in higher dimensions, such as the nonintersection of low-order subspaces. However, after discussing the general theory some simple cases will be examined in detail.

4.3.1. Solution Locus, Common Estimate Locus, and Statistic Locus

Given a sample of n observations y with associated measurements x, the p-parameter model (assumed to be of full rank)

$$E(y) = f(x, \theta)$$

maps the p-dimensional space θ in R_p onto a p-dimensional manifold S in R_n consisting of all possible points $E(y)$. A curve is one-dimensional, a surface two-dimensional, but a p-dimensional manifold can only be imagined, not drawn. S is the solution locus for that model, and for any other formulation of the model that generates the same set of values in some order, by transformation of θ to ϕ. The solution locus for a linear model is a linear subspace. For a given parametrization the coordinate system $\theta_i = \cos t$ may be indicated as "axes" on the solution locus: stable parameter systems generate approximately orthogonal axes, equally spaced for equal increments in each θ_i, while for unstable systems they intersect at very acute angles and may be less equally spaced, discontinuous, or pass more than once through the same point.

The observation y is represented by a point P which does not in general lie in S. Between P and any point of S a distance measure such as the log-likelihood or sum of squares may be defined. The log-likelihood as a squared distance may be computed equally for any other point in R_n, which might belong to some other solution locus S'. Hence, shells of constant distance ("radius") may be defined. For independent, equally weighted, normally distributed errors these shells are spherical and centered at P; for weighted or correlated errors they are concentric ellipses, and for general error laws they may be increasingly nonelliptical and asymmetric. They may be made spherical by transforming to deviance residuals (Section 2.4) relative to P: then the solution loci must also be transformed.

The $(n-1)$-dimensional log-likelihood shell intersects the solution locus in a $(n-p-1)$-dimensional manifold which is the mapping of a likelihood contour in θ. Likelihood contours are fixed in data space; it is only the mapping that changes with parameter transformations. The maximum likelihood estimate $\hat{\theta}$ is at the tangent point θ on the likelihood shell that just touches S. Critical likelihood contours are defined by the shell whose radius exceeds the minimum distance by a prescribed amount.

The class of all points P that share the same maximum likelihood estimate $\hat{\theta}$ lie on an $(n-p)$-dimensional manifold which has been called a *common estimate locus* $C(\hat{\theta})$ (Section 3.1.2). For spherical likelihood shells this is the $(n-p)$-dimensional subspace normal (perpendicular) to the tangent prime to S at $\hat{P}$. In linear models the common estimate loci are parallel and exactly fill R_n, so that to each data point P there is a unique maximum likelihood estimate $\hat{\theta}$. In general, however, the loci are not parallel, intersecting in some parts of R_n and avoiding other parts, dividing R_n into domains where there are 0, 1, 2, or more solutions. These domains were illustrated in Figs. 3.5, 3.6, and 3.8 when it was shown that for data that fit the model well, points P are close to S and have a unique "good" solution, although if S is very curved subsidiary "poor" solutions are possible. If the coordinate system for θ is not 1–1 on S, then of course the common estimate locus will be associated with all the equivalent values of θ.

The class of all points y that share the same value of a statistic, $t(y)$, such as the mean, $\bar{y}$, or sum of squares, or any other function of the data, may be called a *statistic locus*, T. A statistic locus is an $(n-1)$-dimensional manifold. A set of p different statistics generates p such loci which intersect in an $(n-p)$-dimensional manifold U. The locus U passing through P is the set of points sharing the same values of the statistics. Statistics are often proposed as estimators of parameters, so that the intersection of the locus U and the solution locus S is a single point corresponding to parameter values θ^*. A set of statistics such that U coincides with the common estimate locus (θ) through P are known as *sufficient statistics*. In general, an approximate method of estimating θ from sample statistics produces a locus U which is close to the common estimate locus $C(\hat{\theta})$ if P is close to S, but may deviate more and more as P moves away from S and the fit becomes poorer.

To complete the picture, the locus of all points in S corresponding to given values of a function of parameters, $g(\theta)$, may be called a *function locus*, which is a $(p-1)$-dimensional manifold and the relationship of the function locus to P is determined. The object of the experiment is often to estimate $g(\theta)$ given P, and the relationship of the function locus to P is determined by the design of the experiment. Function loci may of course be contours of some alternative parameter system $\phi(\theta)$, but there is no reason why there should be an independent set of p functions of interest; there may be more or fewer.

4.3.2. Examples of the Various Loci in Two Dimensions

The simplest case, $n = 2$ and $p = 1$, may be examined in complete detail. In Fig. 4.17 the following data and models are illustrated:

(a) The data are

$$y_1 = 2.4, \quad x_1 = 1, \quad \text{and} \quad y_2 = 3.7, \quad x_2 = 2,$$

so P is the point (Sections 2.4 and 3.7). Values of x are not explicitly used.

(b) The models:

$f_1(\theta) = \theta x$ leads to the solution locus S_1: $y_2 = 2y_1$.
$f_2(\theta) = \theta^x$ leads to the solution locus S_2: $y_2 = y_1^2$.
$f_3(\theta) = \theta + x/\theta$ leads to the solution locus S_3: $(2y_1 - y_2)(y_2 - y_1) = 1$.

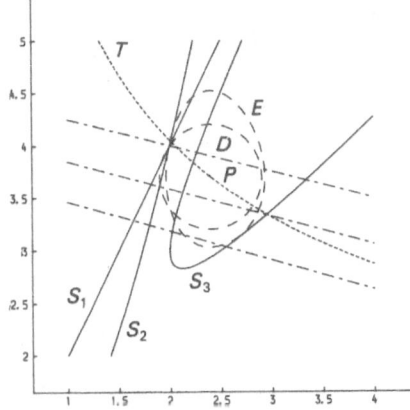

Fig. 4.17. Loci in data space for two observations. (P) Observed data $(2.4, 3.7)$; ———— (S_1, S_2, S_3), solution loci for three models (see text); —————— (D, E), likelihood contours for two error distributions; —·—·—·, common estimate loci for three-parameter values on model f_2; ······ (T), statistic locus for $y_2^2 y_1 = $ constant.

(The third model, f_3, may seem artificial but serves to illustrate typical complexities.)

(c) The likelihood contours:

$$L_1(y): (y_1 - 2.4)^2 + (y_2 - 3.7)^2 = 0.25,$$

a circle representing independent lognormal errors,

$$L_2(y): \log^2\left(\frac{y_1}{2.4}\right) + \log^2\left(\frac{y_2}{3.7}\right) = 0.04,$$

a transformed circle representing independent normal errors.

(d) Common estimate loci for $\theta = 1.8$–2.8 for model $f_2(\theta)$ and likelihood L_1 (normal errors).

(e) Statistic locus T for $y_1 y_2^2 = $ constant, through P. The statistic $y_1 y_2^2$ is an estimator of θ^5.

Parameter contours and function loci are 0-dimensional (i.e., single points) and can be indicated as scale marks on the solution locus. The only matter of importance is how evenly spaced are the scales. On locus S_2, θ is slightly nonlinear, but on S_3, θ is very nonlinear, indicating likely difficulties in applying iterative estimation procedures based on linear approximations. Transformations to $\phi(\theta)$ are likely to be stable if the scale is more even. If the solution locus is locally nearly linear, intersection with a family of statistics loci may provide a suitable system. However, we have to work with functions of θ, not $\hat{y}$, and so rather than choose the reparametrization $\phi = \theta + \theta^2$ for S_2, a function such as $\phi = \theta^{1.5}$ could be used to have similar effect.

To investigate the effects of using lognormal errors, all the loci may be redrawn on the scale of $\log y_1$ and $\log y_2$. Likelihood contour L_2 becomes a circle and S_1 and S_2 become straight lines. The statistic locus T also becomes a straight line and may be seen to coincide with the common estimate locus for f_2 through $\theta = 2.01$. This illustrates how an approximate estimation method may become exact under different error assumptions.

Model f_3 remains difficult to interpret, with two possible solutions, unless some convention is adopted that restricts the range of θ.

4.3.3. Three Data Points and Two Parameters

The representation of three data points may be made in several ways: as a solid model using plaster, wire or other materials; as a perspective or oblique drawing of such a model; or as an orthogonal projection, horizontally as contours or vertically as cross-sectional profiles.

The solution locus for a one-parameter model such as $E(y) = \theta^x$ is a spatial curve, in general not lying in any plane, whose shape may be represented by bending stout wire, following the values of its parametric coordinates such as $(\theta, \theta^2, \theta^3)$ which is a twisted cubic, the intersection of the quadrics $y_1 y_2 = $

y_3 and $y_2^2 = y_1 y_3$. Common estimate loci are a pencil of planes such as

$$y_1 + 2\theta y_2 + 3\theta^2 y_3 = \theta + 2\theta^3 + 3\theta^5.$$

The solution locus for a two-parameter model is a surface, $y_i = f_i(\theta_1, \theta_2)$. To plot cross-sectional profiles of y_3 as a function of y_1 and y_2, it may be necessary to use numerical interpolation of find given values of y_3, if no algebraic simplification is possible. For example, the two-compartment model (see Example 3.1)

$$y_i = \frac{\theta_1 \exp(-\theta_2 x_i) - \theta_2 \exp(-\theta_1 x_i)}{\theta_1 - \theta_2}$$

cannot be solved explicitly by eliminating θ_1 and θ_2, but particular values of θ_1 and θ_2 that generate a given value of y_3 may be found and corresponding values of y_1 and y_2 plotted (Fig. 4.18). This diagram may be converted into a solid model by cutting the profile onto pieces of thick card which are then assembled at equally spaced intervals like a flight of stairs; the model is completed by filling the spaces with plaster. The parameter grid may now be drawn on the surface choosing suitable values of θ_1 and θ_2. Alternative parametrizations may be represented, and likelihood contours for given data points and error distributions. Such models are of educational rather than practical value, as are wire mesh representations which are even more complicated to construct. The two-dimensional representation (Fig. 4.18) can still be used directly, by plotting in the space of (y_1, y_2) the contours of y_3, θ_1, θ_2, and $L(\theta_1, \theta_2)$ in distinguishable form. The shapes of the likelihood contours are distorted by the projection, but this distortion can be lessened by suitable choice of y_1 and y_2, if choice exists. Functions such as extra predicted values, y_4, may also be displayed as contours on this diagram (Fig. 4.19).

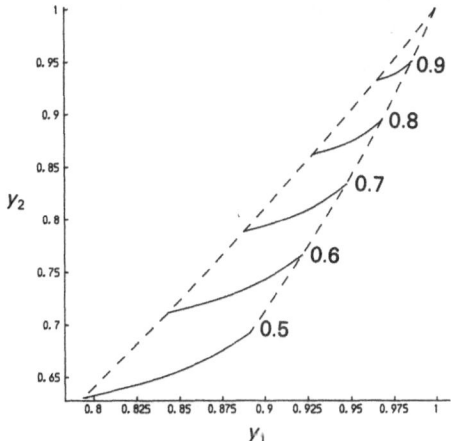

Fig. 4.18. Solution locus contours for three observations at equal intervals $(y_1 \ y_2 \ y_3)$. Plots of (y_1, y_2) for given values of y_3 are shown for the two-compartment model (see text). The broken lines mark the limits of the interpretable parameter region.

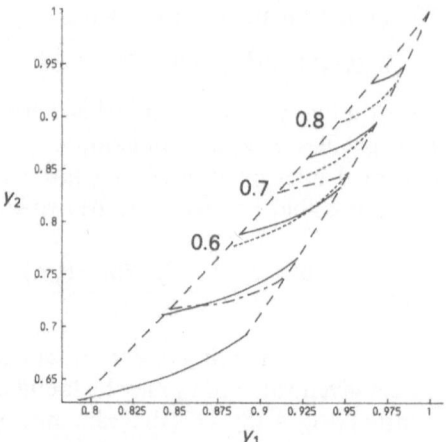

Fig. 4.19. The picture in Fig. 4.18 with superimposition of likelihood contours and function loci. $-\cdot-\cdot-\cdot$, two parts of the same likelihood contour; $\cdots\cdots$, function loci for predictions at $x = 4$ ($y_4 = 0.6, 0.7, 0.8$).

Common estimate loci are one-dimensional and their portrayal is confusing on a plane projection. The question of how they intersect, or leave regions unrepresented, is difficult to quantify in more than two dimensions, except in the situations described in Section 3.1.2, where linear parameters are eliminated, and ratios or differences are plotted.

4.4. Plots in Parameter Space

Parameter space is a subset of p-dimensional space consisting of the set of values of θ which are allowable in a particular model. The values may sometimes be constrained, say to the positive quadrant or the unit square, or in case of symmetry, to the triangular region where, say, $\theta_1 \le \theta_2$. To interpret models most easily it is best to use stable parameters, if possible, with suitable pseudomodels to extend the interpretation of the parameters beyond those corresponding to valid ranges of θ.

One- and two-parameter models may be studied completely using graphs and contours. For three or more parameters several plots may be needed, for example:

(a) A series of contours with respect to θ_1 and θ_2, using a grid of values of θ_3, θ_4, etc., as in Fig. 4.20.

(b) A contour diagram for each pair of parameters in turn, usually centred at $\hat{\theta}$. The plots may be displayed as a triangular array such as

$$\theta_2 \vee \theta_1,$$

$$\theta_3 \vee \theta_1, \quad \theta_3 \vee \theta_2,$$

$$\theta_4 \vee \theta_1, \quad \theta_4 \vee \theta_2, \quad \theta_4 \vee \theta_3,$$

as in Fig. 4.21.

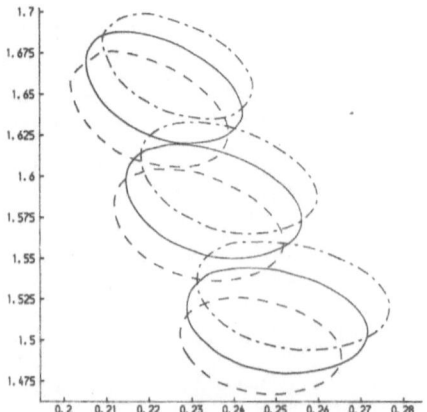

Fig. 4.20. Plotting a four-dimensional likelihood surface (1). The model is the logistic curve fitted to the data of Fig. 4.3. The same contour level is shown in (θ_1, θ_2) space for the fitted values of (θ_3, θ_4) and for equally spaced steps either side of the fitted values.

(c) Cross-sectional profiles between any pair of points $\hat{\theta}_a$ and $\hat{\theta}_b$, or a series of such profiles. The directions chosen may be parallel to the axes, or oblique. These plots are particularly useful for revealing local optima.

(d) Oblique coordinates, for use when parameters are highly correlated, and when constraints prevent the use of rectangular coordinates. Values are computed for a diagonal lattice defined by a parallelogram corresponding to three corner values $\hat{\theta}_a$, $\hat{\theta}_b$, $\hat{\theta}_c$: the parallelogram is projected onto a rectangle, and narrow ellipses become much wider and easily drawn.

The functions to be discussed are as follows:

(1) The log-likelihood or sums of squares function, $L(\hat{\theta})$.
(2) Functions related to $L(\theta)$, such as first derivatives, $L'(\theta)$, or discrepancy functions, $L(\theta) - Q(\theta)$, where Q is the quadratic approximation to L at $\hat{\theta}$.
(3) Conditional log-likelihoods, $L(\theta, \hat{\beta}(\theta))$, in which linear parameters $\beta(\theta)$ have been eliminated analytically. The functions $\hat{\beta}(\theta)$ may also be displayed.
(4) Function plots $g(\theta)$ of quantities of interest, including potential transformations $\phi(\theta)$.
(5) *Parameter loci* or plots of particular solutions of $f_i(\theta) = y_i$, or of compound expressions $h(f_i, f_j) = h(y_i, y_j)$ such as differences or ratios which eliminate linear parameters.
(6) Convergence contours, showing the behavior of a given algorithm on a given problem.

The relationships between these plots and those described in the previous section are of interest.

Likelihood contours are mapped from the fixed likelihood contours on the solution locus using the parameter system $(\hat{\theta})$. Functions of parameters do not

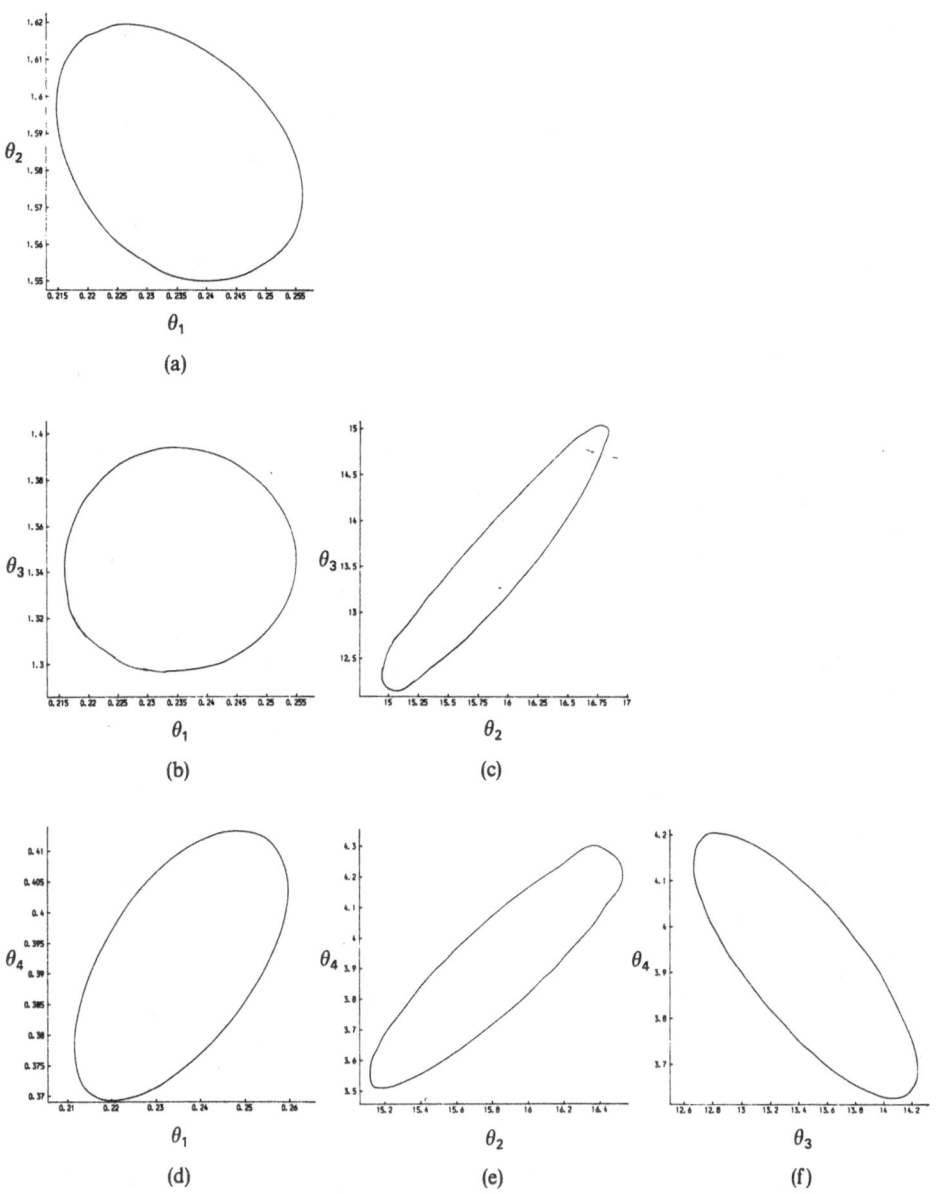

Fig. 4.21. Plotting a four-dimensional likelihood surface (b). The same contour level and model as in Fig. 4.20 displayed in pairs of dimensions, arranged in triangular form.

necessarily correspond to functions of data, and in nonlinear models there is clearly a distinction between $g(f(\theta) = c)$, the set of points on the solution locus that belong to $g(y) = c$, and $g(\theta) = k$, a direct function of parameters. Parameter loci correspond to intersections of the solution locus with linear subspaces through the data point, such as $y_i = Y_i$, and are therefore particular cases of functions of data.

4.4.1. Likelihood Contour Plots

There have been examples of contour plots in previous chapters (Figs. 2.1, 2.2, 3.9, 3.10), their purposes being to show the existence and uniqueness, or otherwise, of maximum likelihood estimates (MLEs), and the shape of likelihood-based confidence regions. Stable parameters are justified by the appearance of compact elliptical contours. A priori, the region of parameter space chosen for plotting must be guessed at; a posteriori, a suitable region that gives reasonable diagrams is centered at the MLE, the range of each parameter being about four standard errors each way, and the contour intervals chosen to coincide with critical values of the likelihood ratio tests. Such plots may be generated automatically, though care must be taken to avoid nonfeasible regions of parameter space, if necessary by defining pseudomodels or by restricting the ranges of particular parameters.

4.4.2. Derivatives and Discrepancy Plots

Contours of derivatives may be generated from differences between $L(\theta)$ and $L(\theta + \delta\theta)$. For linear models the contours should be straight lines, evenly spaced. The contours of $\partial L/\partial \theta = 0$ for each θ_i intersect at the MLEs, and if the parameter estimates are uncorrelated they intersect at right angles.

Discrepancy plots (Fig. 2.7) are a graphic way of demonstrating the extent to which a quadratic approximation is valid. By subtracting the quadratic terms of the Taylor expression of $L(\theta)$, the remainder function is dominated by cubic terms. The cubic $C(\theta) = 0$ either has one or three real roots corresponding to one or three linear contours through $\hat{\theta}$; if there is only one real root, the discrepancy function is positive on one side of the line, negative on the other, indicating asymmetry maximized at right angles to this line, with contours at L too close together where C is positive, too far apart where C is negative; when there are three real roots a pattern of alternate positive and negative regions indicates curvature relative to the expected elliptical contours, although asymmetry will also be involved. By superimposing the discrepancy plot on the likelihood plot, the maximum extent of the discrepancy on a critical contour may be estimated.

4.4.3. Elimination of Separable Linear Parameters

Parameters that may be estimated analytically using the equation $\partial L/\partial \theta_i = 0$ are known as separable *linear* parameters (Section 5.3), β_1, and their estimates $\hat{\beta}(\theta)$ are functions of the remaining parameters. Values of $\hat{\beta}$ may be plotted as contours, and in cases where the value of $\hat{\beta}$ must lie in a certain interval, such as $\hat{\beta} \geq 0$, the boundary contour imposes a further constraint on θ.

The value of the log-likelihood after substituting $\hat{\beta}(\theta)$ is $L(\theta, \hat{\beta})$ which may be plotted for the nonlinear parameters only. Critical contours of $L(\theta, \hat{\beta})$ define a likelihood-based confidence region for θ. To obtain confidence regions for β the full parameter set must be studied. The best known examples of linear parameters are models of the form $E(y) = \sum \beta_i f_i(\theta)$ with normally distributed errors in which for given θ, the β parameters are estimated by linear regression. Less well known are scale parameters in Poisson and gamma distributed variables such as β, where

$$E(y) = \beta f(\theta)$$

and in the Poisson model,

$$\hat{\beta} = \frac{\sum y}{\sum f},$$

and in the gamma model,

$$\hat{\beta} = \frac{\sum (y/f)}{n}.$$

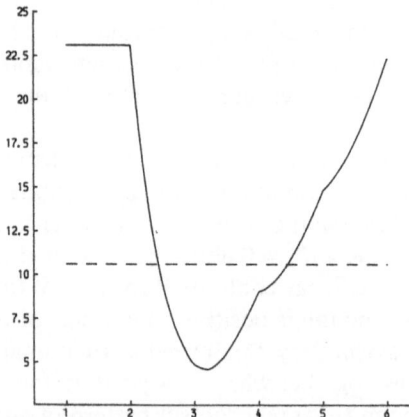

Fig. 4.22. Residual sum of squares plotted as a function of the break-point parameter in the linear spline model. Data adapted from Fig. 2.1 by replacing y_2 by 9 instead of 12. A critical level of the residual sum of squares is indicated, the intercepts giving confidence limits to the break-point estimate.

In other case the normal equations may only be solved numerically, so that elimination is not a practical procedure.

An important example is in fitting spline functions with θ giving the positions of the nodes or changes of phase (see Section 7.2.3). Given θ the model is linear, but the conditional log-likelihood function is discontinuous in slope as a particular node θ_j passes through a data value of x. It is therefore important to plot the function to confirm the uniqueness of the solution, and to obtain confidence limits, which are very different to those predicted from the quadratic approximation. Figure 4.22 shows the plot for a single node for two straight lines fitted to the data of Fig. 2.4, with $y_2 = 9$ instead of 12.

4.4.4. Function Plots and Potential Transformations

The discussion of parameters in Section 2.1 distinguished three roles of parametrization: stable parameters for fitting models, defining parameters for evaluating expectations, and interpretable parameters for estimating useful quantities.

The representation by stable parameters ϕ makes it necessary to have some means of recovering information about $\theta(\phi)$, the defining parameters, and of $g(\phi)$, the interpretive parameters. Contours of θ and g may be drawn in ϕ space and superimposed on a plot of the critical likelihood contour, $L_c(\phi)$. The contours of g that touch the L_c contours correspond to the greatest and least value of g for ϕ values within the confidence region. If contours of g are approximately parallel straight lines, equally spread, the function is locally linear and linear approximations may be adequate. In more extreme cases the contours are curved and unequally spaced, and may touch the L_c contours in

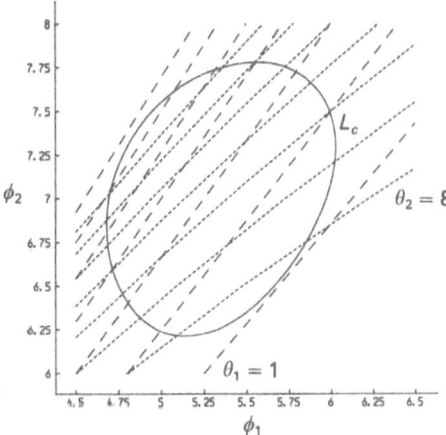

Fig. 4.23. Plot of parameter transformations, showing relationship between defining parameters and stable parameters for the Michaelis–Menten example (see Fig. 2.1(b)).

more than one place, indicating discontinuous confidence intervals for g, especially for functions that may become infinite.

In Fig. 4.23 the Michaelis–Menten parameters θ_1 and θ_2 are plotted against the stable parameters ϕ_1 and ϕ_2 of Fig. 2.1. The limiting tangential contours are clearly shown, together with the asymmetry of the spacing and the very acute angles at the intersections between the two systems. Note that if the contour L_c had crossed the line $\phi_2 = 2\phi_1$ the confidence intervals for θ_1 and θ_2 would be discontinuous.

The main purpose of these plots is to help to explain difficulties that may arise in computing exact confidence limits for functions of parameters. When the likelihood contours indicate that parameters are unstable, plots of potential transformations may help to suggest suitable choices empirically, irrespective of theoretical justifications for their use.

4.4.5. Parameter Loci

The exact solutions of the equations $f_i(\theta) = y_i$ are sets of θ values for which a particular observation fits the model exactly, with zero residual. These solutions lie on a locus of dimension $(p - 1)$, which may be called *parameter locus* which is a special case of the family of loci of constant residuals, a_i.

The intersection of a set of k parameter loci is a locus of dimension $(p - k)$, and if $km = p$ the locus is a point. The point may or may not be unique, or may be imaginary: curved loci any intersect more than once or not at all, repeated observations at the same design point give parallel parameter loci. Repeated observations should be combined into a single observation at the weighted mean response, when studying parameter loci.

Curves for which the deviance residuals $e_i = 0$ are the parameter loci and the log-likelihood function is the sum of components e_i^2. The graph of all parameter loci therefore indicates the structure of the likelihood function, depending on the nature of the error distribution.

If we assume that there are n distinct design points, then any set of p observations from which all p parameters may be determined is a potential solution, in particular, if all other points are given zero weight. These solution points in parameter space may be called *nodes*. The set of parameter loci divides parameter space into domains, and within any domain the pattern of positive and negative residuals is the same. If we consider solutions lying on any line in parameter space, the log-likelihood must decrease monotonely towards the center in the two outer domains, since all the component likelihoods decrease towards their respective points of intersection with their parameter loci. Hence solutions must lie in some inner domain where components that are increasing balance those that are decreasing.

To illustrate the argument, consider the data of Table 2.1 fitted to the rectangular hyperbola through the origin. Using stable parameters ϕ_1 and ϕ_2, defined ordinates at $x = 3$ and $x = 6$, the model may be written

$$E(y) = \frac{\phi_1 \phi_2 x}{(2\phi_1 - \phi_2)x + 6(\phi_2 - \phi_1)}.$$

Parameter loci obtained by substituting data values of y in this formula give hyperbolic curves relating ϕ_1 and ϕ_2, namely

$$\phi_2 = \frac{2\phi_1 y_i(x_i - 3)}{\phi_1 x_i + y_i(x_i - 6)}, \qquad i = 1, \dots, 7,$$

which are illustrated in Fig. 4.24. Curves 3 and 6 are of course straight lines and the seven curves intersect at 21 nodes. The shaded inner region is a polygonal area within which all solutions must lie, whatever weighting is employed, although solutions may be on the boundary or at nodes. Decreasing a weight to zero is equivalent to removing a locus, and it is easy to identify outliers as points whose removal would make the remaining nodes more compact a set. For example, the removal of point 6 would have the largest effect. Note that nodes (4, 5), (4, 6), and (5, 6) are below the diagonal $\phi_2 = \phi_1$ and correspond to a concave hyperbola with asymptote to the left. The diagram, if plotted in terms of parameters θ_1 and θ_2 of

$$E(y) = \frac{\theta_1 x}{x + \theta_2},$$

would be more confusing: the curves

$$\theta_2 = \frac{\theta_1 x_i}{y_i} - x_i$$

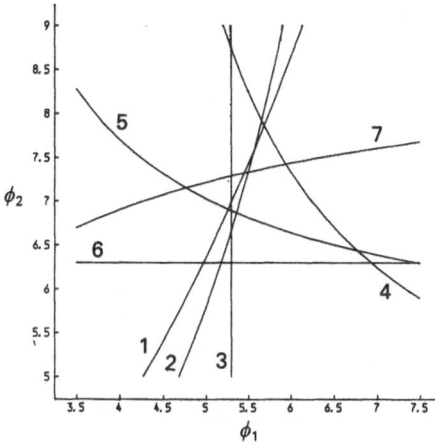

Fig. 4.24. Parameter loci for the Michaelis–Menten example (Table 2.1) showing parameters that fit each data point exactly.

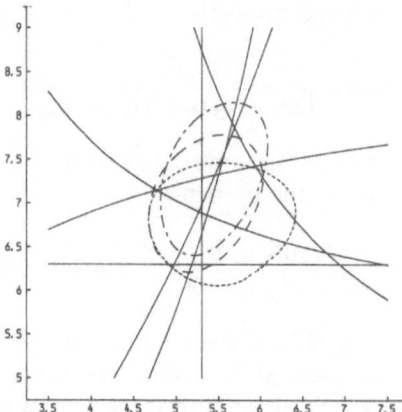

Fig. 4.25. Likelihood contours using different weights for the data of Fig. 4.24, showing the effect of different data points. ‒ ‒ ‒ ‒ ‒ ‒ ‒ ‒ ‒, equal weighting; ‒·‒·‒·‒·, increased weight to 4, 5, and 6; ·······, decreased weight to 5 and 6.

are nearly parallel straight lines, not all of which intersect in the positive quadrant. In this model, that the lines are straight shows that each pair of parameter loci intersects in a single node, because the loci correspond after transformation of parameters. This is not true in general, but is a property of ratios of polynomials. It does not prove that the likelihood function has a unique optimum, because the way in which components of likelihood combine may produce local optima.

The log-likelihood contours (critical sums of squares), for various weighted normal errors, are illustrated in Fig. 4.25. Up-weighting points 4, 5, and 6 moves the solution towards the triangle where their loci cross, while down-weighting points 5 and 6 moves the solution upwards.

4.4.6. Parameter Loci for the Two-Compartment Model

The two-compartment model

$$E(y) = \frac{\theta_1 \exp(-\theta_2 x) - \theta_2 \exp(-\theta_1 x)}{\theta_1 - \theta_2}$$

has been discussed in Section 3.1.1, but it is instructive to consider it in terms of parameter loci. It is best illustrated in the space of two ordinates, $\phi_1 = E(y(x_1))$ and $\phi_2 = E(y(x_2))$, which eliminates the essential ambiguity between θ_1 and θ_2.

In Fig. 4.26 parameter loci are illustrated for the data:

$x =$	1	2	3	4	5	6
$y =$	0.92	0.81	0.76	0.68	0.61	0.59

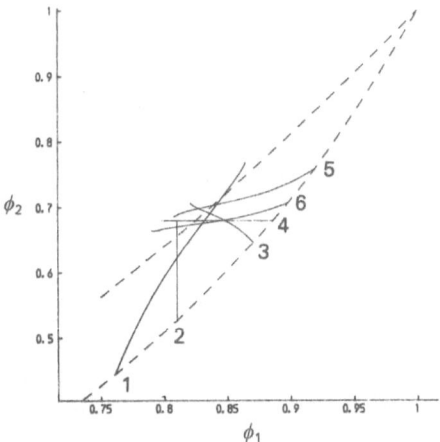

Fig. 4.26. Parameter loci for the two-compartment model (see text).

in terms of $\phi_1 = E(y(2))$ and $\phi_2 = E(y(4))$, with values found by interpolation. Note that the loci are only defined for real θ values in the region between the two limiting curves derived from solutions of

$$E(y) = \exp(-\theta x),$$

and

$$E(y) = (1 + \theta x)\exp(-\theta x).$$

Since the loci do not all intersect in this region it is clear that there will be problems in fitting these data, the MLEs tending to lie outside the valid region. Locus 2 has fewest intersections with other loci, and could be considered an outlier (if the model is true).

To extend the loci until they cross, it is necessary to find suitable pseudomodels that define values in the regions adjoining the valid region. The model

$$E(y) = \frac{\theta_1 \exp(-\theta_2 x) + \theta_2 \exp(-\theta_1 x)}{\theta_1 + \theta_2}$$

has the necessary continuity properties, but since the solutions for $\theta_2 = 0$ and $\theta_2 = \theta_1$ are identical, the parameter loci do not extend very far before they reverse direction. This model does however allow more of the likelihood function to be drawn. On the other boundary of the valid region the extension is provided by the model

$$E(y) = \exp(-\theta_1 x)\cos(\theta_2 x) + \left(\frac{\theta_1}{\theta_2}\right)\sin(\theta_2 x).$$

4.4.7. Representation of Three or More Parameters

Parameter loci (Section 4.4.5) in three or more dimensions are not easily described in detail, nor is it worth the computing effort to construct them.

Instead it is possible to gain some insight by reducing the dimensionality of the problem, either by regarding well-estimated parameters as fixed (and plotting solution loci for the two most awkward parameters), or by elimination parameters from sets of $q + 1$ observations. The first idea needs no further discussion, but the second idea can be very useful.

The parameters most easily eliminated are linear parameters, eliminated by taking differences or ratios of data values, or solving linear equations. For example, to study the general double exponential curve

$$E(y) = \beta_0 + \beta_1 \exp(-\theta_1 x) + \beta_2 \exp(-\theta_2 x),$$

sets of four data points are required to eliminate the β parameters, to illustrate solution loci in (θ_1, θ_2) space. Since θ_1 and θ_2 are symmetric it is preferable to transform, say to (ϕ_1, ϕ_2) space, where θ_1 and θ_2 are roots of the quadratic equation

$$t^2 - 2\phi_1 t + \phi_2 = 0.$$

Clearly, we do not wish to solve for every set of four data points: widely spaced sets are more likely to be informative, and provided every data point is included at least once, the characteristics of the problem may be inferred. For four equally spaced points the loci are straight lines.

$$\phi_2(y_2 - y_1) - \phi_1(y_3 - y_2) + (y_4 - y_3) = 0.$$

Data given by Lipton and McGilchrist (1963) are as follows:

x	0	1	2	3	4	5	6	7	8	9	10	11
y	9.21	6.01	5.11	4.22	3.98	3.42	3.60	3.38	3.48	3.17	3.18	3.11

and suitable loci to illustrate are those for $x = (0, 3, 6, 9)$, $(1, 4, 7, 10)$, and $(2, 5, 8, 11)$. The authors found that the solution did not lie in the real plane of (θ_1, θ_2) and gave a pseudomodel solution. It is therefore not surprising that the loci in (ϕ_1, ϕ_2) space (illustrated in Fig. 4.27), do not enter the region corresponding to the real value of θ. Note that this method does not indicate which points are causing the difficulty: changes in any of the four points will alter the lines in some way, and the effect on the residual sum of squares depends on the position on the curve.

An alternative treatment is to regard the values for $x = 0, 1$, and 2 as fixed, because their effective replication is close to 1 in this case. The loci for each point y_3, y_4 onwards may then be drawn, but being in general nonlinear involves greater computing effort. The first three are however easily constructed, and show again that the loci intersect outside the feasible region. To plot more intractable loci it is simplest to use a contouring routine to solve the intrinsic equation $\lambda(\phi_1, \phi_2) = 0$.

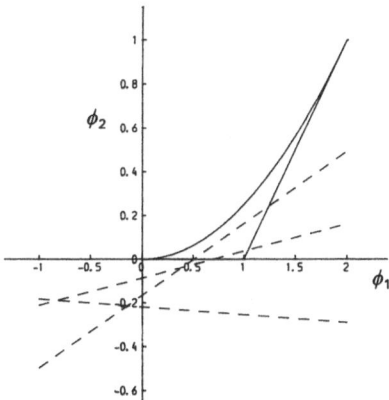

Fig. 4.27. Parameter loci for nonlinear parameters in a separable model using groups of data values (see text).

4.4.8. Convergence Loci

Convergence loci are graphical aids to studying how rapidly and safely a method of optimization applies to a particular problem. In one-parameter models a plot of the next iteration $\theta^{(1)}$ as a function of the current iteration $\theta^{(0)}$ will reveal sufficient information. The curve crosses the 45° line $\theta^{(0)} = \theta^{(1)}$ at the solution, and for quadratically convergent methods the gradient $\partial\theta^{(1)}/\partial\theta^{(0)}$ is zero at the solution. If the gradient is nonzero the method is only linearly convergent and approaches the solution in steps in approximately geometrical progression, which is the justification for Aitken's d^2 technique. Figure 4.28 illustrates the progress of two different algorithms applied to the same problem: the first converges quadratically for good initial estimates but diverges for poorer starts, the second converges only linearly but will converge for all points in a wide range. Plots of second or third iterations may also be made.

The representation of convergence with respect to two parameters can be shown in several ways. For example, a rectangular grid of values $(\theta_1^{(0)}, \theta_2^{(0)})$ gives rise to a corresponding array of values $(\theta_1^{(1)}, \theta_2^{(2)})$, the lines on the grid transform to curves, which may cross themselves, collapse to a point, or become extended. The two grids are plotted on the same diagram, which is constructed by recording the iterations resulting from each starting value.

At alternative approach is to contour the log-likelihood values achieved after a given number of iterations. If a particular level is regarded as "converged," its contour, C_0 say, represents acceptable solutions. The contour C_1 defines a set of points such that the results of one iteration lie on C_0, and it is assumed that points within C_1 converge to points with C_0. Ideally, a set of nested contours C_1, C_2, C_3, etc. is defined, so that one can judge the behavior of an algorithm in terms of the region of acceptable starting values.

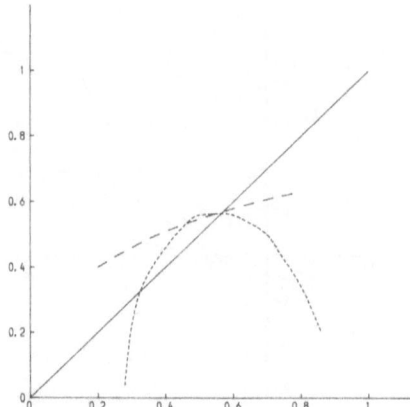

Fig. 4.28. Plot of $\theta^{(1)}$ against $\theta^{(0)}$ to illustrate convergence behavior of algorithms. $----$ linear convergence, $\cdots\cdots$ quadratic convergence.

If parameters are transformed, corresponding contours C_1^*, C_2^*, C_3^* may be drawn in the original parameter space to show how the transformation affects convergence.

These studies are particularly useful for simple algorithms, such as pure Gauss–Newton or Newton–Raphson, which use no logical modifications to assist convergence. In practice, algorithms behave in a discontinuous manner if certain conditions are true or false; for example, a maximum step length

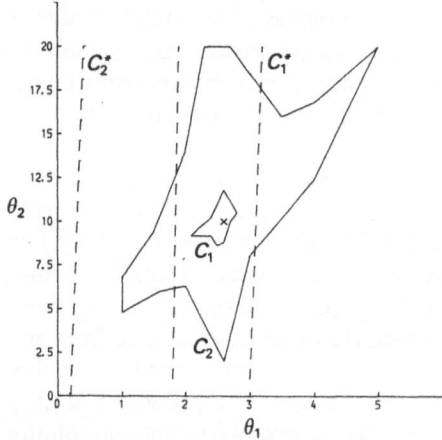

Fig. 4.29. Plot of convergence loci in two dimensions to illustrate the advantage of stable transformation. ———— C_1, C_2, convergence in one or two iterations using the Gauss–Newton method for the Michaelis–Menten problem (Table 2.1); $-------C_1^*$, C_2^*, the value using stable parameters.

may prevent divergence by putting a limit on $\partial \theta^{(1)}/\partial \theta^{(0)}$ in the one-parameter case.

Figure 4.29 illustrates convergence loci for the Michaelis–Menten example (Fig. 4.1) with and without transformation, using the criterion for convergences RSS < 0.931. The convergence contours C_1 and C_2 for Gauss–Newton iterations using parameters θ_1 and θ_2 are shown, illustrating the importance of a good initial estimate of θ_1. Using the transformation to (θ_1, θ_2) space, contours C_1^* and C_2^* are for the same problem expressed in terms of the original parameters, showing that an initial estimate of θ_1 in the range $(1.8, 3.2)$ will converge in one iteration, and any initial estimate in the range $\theta_1 > 0.25$ will converge in two iterations. The convergence contours may tend to a limiting contour C_0 outside which there is divergence.

CHAPTER 5

Computing Methods for Nonlinear Modeling

5.1. Computer Programs, Libraries, and Packages

Computer packages for nonlinear modeling may be classified as follows:

(1) Customized routines with local conventions, for particular curves, distributions or other models, such as the logistic curve, the Weibull distribution, or the probit line for quantal response in bioassay.
(2) General optimization algorithms supplied by subroutine libraries: users must organize their data, output, initial values, and constants for the optimization algorithm.
(3) General statistical algorithms such as iteratively weighted least squares, which handle a subclass of nonlinear models in a standard manner, such as GLIM.
(4) General statistical packages which include options for general model fitting and optimization, such as GENSTAT, SAS, or BMDP.
(5) Specialized nonlinear modeling packages with facilities for both standard models and general user-defined models, such as the maximum likelihood program MLP (Chapter 7).

The customized routine (1) is often adequate in an environment where a single model is to be fitted repetitively; input and output may be planned to suit local needs. A drawback is the inflexibility of the method to minor changes in the problem. Attention to the efficiency of solution can make dramatic differences to the speed and reliability of such routines.

A general optimization algorithm (2) offers a solution to the numerical analysis problem of finding optima of a general function. There is no suggestion that the function may be modified in any way, although the behavior of the algorithms may be considered in terms of empirical transformations based on trial function values. These algorithms are most appropriate for solving unique problems when the work required to find the best formulation of the problem is not justified.

General statistical algorithms (3) work well for some problems, poorly for others, and not at all for some important applications. This mean that the user's choice of model may be unduly restricted by the constraints of the algorithm. For example, it is easy to fit an exponential curve with known asymptote using GLIM, but not so easy when the asymptote is unknown although the "offset" facility may be of use.

General statistical packages (4) provide links between standard data descriptions and optimization algorithms, enabling the user to define arbitrary models and to use the output and graphical applications provided by the package. The user may still have difficulty with the appropriate formulation of the problem; some packages assume that it is sufficient always to fit by least squares rather than by maximum likelihood.

General modeling packages (5), such as MLP, attempt to provide as wide a class as possible of the commonly occurring applications as the standard models, thus making the fitting of these models available to nonstatisticians who would have difficulty in appreciating the problems associated with nonlinear modeling. Specialized forms of output appropriate to each standard model may be incorporated, and methods for comparison of models and data sets, such as parallel model analysis, may be provided. A full range of diagnostic aids for general nonlinear modeling may be provided. The question of whether systems such as the MLP should be subsets of larger packages, should have interfaces with other packages, or should incorporate such other features as may normally be required in a nonlinear modeling application, depends on the computing environment. Some facilities of MLP are available in GENSTAT 5 (Lane *et al.*, 1987).

5.2. Requirements for Fitting Nonlinear Models

The complete process of model fitting may be described as follows:

(1) *Data input and description*
The size of the data array, the nature of the variates, the values of any input constants, the structure of parallel data sets (Section 1.1.2), or factor variates, must be provided. Facilities for displaying data (Section 4.2), making transformations and constructing extra variates such as weights are also required.

(2) *Model selection or definition*
The observation variate, the expectation variate as a function of parameters, the error distribution (including any weight variate) must be defined (Section 1.1). The nonlinear relationship may be defined by name (for standard models) or by formula (for general models).

(3) *Data screening and preliminary analysis*
The data must first be checked: that the variates are appropriate and that the sample size is sufficient for estimating parameters in the model. Sample statistics such as the ranges, means, variances, and correlations may be computed. These quantities will be needed later both to provide working constants for parameter transformations and to provide initial estimates of parameters. There may be sufficient information at this stage to reject certain data sets if the boundary region is well defined: for example, the parameter k in the negative binomial distribution (3.1.16) has a negative maximum likelihood estimate (MLE) if the sample mean exceeds the sample variance (see, e.g., Ross and Preece, 1985).

(4) *Choice of methods of fitting*
Where direct methods of fitting exist, they should be used; there is little merit in relying on a single algorithm to solve all problems. The solution strategy depends on the information available and the complexity of the problem. Some problems may need no transformation, others may need a mixed strategy depending on the results of initial trials. Linear parameters may be handled separately with advantage (see 5.4).

(5) *Assignment of working constant and initial values*
Data-dependent transformations (Section 2.4.3) require working constants such as working means and scale factors. The optimum values of these may not be known in advance. Provisional values are assigned which may be adequate to obtain a fit. Initial values are derived either from the data statistics, from mid-values of the parameter range, or from the best of a few trial values. If convergence is not achieved within a few iterations the working constants may be revised for a second attempt.

(6) *Fitting the model and estimating the dispersion matrix*
If the method does not converge, it may be possible to redefine the problem and try again. If bounds are violated limiting cases may be fitted.

(7) *Recovery of defining parameters*
If the inverse transformation corresponds to a pseudomodel (Section 3.1.4), the parameters may be output with a suitable warning, or else the limiting acceptable case given. When several data sets are compared in an analysis of parallelism (Section 1.1.2), it is necessary to use the full pseudomodel solution, to obtain the appropriate log-likelihood.

(8) *Output of functions of parameters*
Expectations, slopes, and other appropriate functions may be computed together with their standard errors, using the linear approximation formula (Section 3.2).

(9) *Output requiring further search algorithms*
Methods for computing confidence limits (Section 3.2), inverse interpolation, and *a posteriori* stable transformations (Section 2.3) require knowledge of the fitted model, and may be called at this stage (Section 1.2.2).

(10) *Combination of results*
Given a sequence of data sets or model requests the results of intermediate stages must be stored so that the combined analysis has all available information. The results may be set out as an analysis of variance or deviance (Section 1.2.2).

5.2.1. Example of the Fitting Procedure: Logistic Curves

To illustrate the above process, consider a routine to fit the logistic curve

$$E(y) = \beta_0 + \frac{\beta_1}{1 + \exp(-\theta_1(x - \theta_2))} \qquad (5.2.1)$$

to n data pairs (x, y), with possible weights w.

(1) When the standard error is assumed to be proportional to the expectation, a provisional weight variate proportional to $1/y^2$ is required. This may be revised when fitted values of y have been estimated.
(2) Provision must be made for the constrained model, $\beta_0 = 0$.
(3) If $n < 4$ there are insufficient data values ($n < 3$ if $\beta_0 = 0$), and in $n = 4$ the fit will be exact and there will be no degrees of freedom for error. If some weights are zero or x values are repeated there may also be insufficient data values. To test the latter it would be necessary to sort the data by x and check for repetitions.
(4) Parameters β_0 and β_1 are linear if θ is known, and so the method of fitting will be to optimize the residual sum of squares as a function of θ or some transformation of θ. A suitable transformation based on ratios of ordinates is as follows.
   Let $x_1 = m - s$, $x_2 = m$, $x_3 = m + s$, be three equally spaced values of x. Then if b_0 is a provisional estimate of β_0 the ratios

$$\phi_1 = \frac{f(x_1) - b_0}{f(x_2) - b_0}, \qquad \phi_2 = \frac{f(x_2) - b_0}{f(x_3) - b_0}, \qquad (5.2.2)$$

may be used as parameters, being related to θ_1 and θ_2 by the equations

$$r = \frac{\phi_1(1 - \phi_2)}{1 - \phi_1},$$

$$k = \frac{(1 - \phi_1)(1 - \phi_2)}{\phi_2 - \phi_1},$$

$$\theta_1 = -\frac{\log(r)}{s},$$

$$\theta_2 = m + \frac{\log(k)}{\theta_1}, \tag{5.2.3}$$

defined over the triangular region $0 < \phi_1 < \phi_2 < 1$.
    If $\phi_1 = \phi_2$, fit the exponential curve

$$E(y) = \beta_0 + \beta_1 \exp(\theta_1 x). \tag{5.2.4}$$

If $\phi_1 > \phi_2$, fit the "catastrophic" curve

$$E(y) = \beta_0 + \frac{\beta_1}{1 - \exp(-\theta_1(x - \theta_2))}, \tag{5.2.5}$$

where $x = \theta_2$ is a vertical asymptote. If ϕ_1 or ϕ_2 lie outside the range $(0, 1)$ the curve is not monotone, and the limiting case is a step function

$$E(y) = \beta_0, \qquad x < \theta_2,$$

$$= \beta_0 + \beta_1, \qquad x > \theta_2.$$

Note that $f(\beta_0, \beta_1, \theta_1, \theta_2) = f(\beta_0 + \beta_1, -\beta_1, -\theta_1, \theta_2)$ because the curve is symmetric about θ_2. By convention β_1 is positive so that θ_1 is positive for monotone increasing curves and negative for monotone decreasing curves, achieved by giving a sign to s.
(5) A minor problem is to estimate m and s so that the residual sum of squares function is easy to optimize. The key assumption is that the shape of the curve suggested by the plotted data bears some resemblance to the logistic curve, so that values of m and s derived from the data could produce a stable parameter system. It is not sufficient to look at the range or standard deviation of x because for some distributions the chosen ordinates will be close to the asymptotes. Instead the following procedure is usually effective:
(a) Find the maximum, minimum and range of y.
(b) Find data points close to

$$y \min + c \cdot y \text{ range}, \qquad c = 0.1, 0.5, \text{ and } 0.9,$$

and let the corresponding values of x be x_1, x_5, and x_9. Then choose

$$m = x_5, \qquad s = \frac{x_9 - x_1}{3}, \qquad \phi_1 = 0.4, \qquad \phi_2 = 0.625,$$

as initial settings.
(6) Optimize with respect to ϕ, with the constraints $0 < \phi_1, \phi_2 < 1$. If an interior solution is found, then if $\phi_1 < \phi_2$, it corresponds to the required logistic curve. If $\phi_1 \geq \phi_2$ it corresponds to a pseudomodel or to the limiting exponential. If both ϕ_1 and ϕ_2 tend to the upper limit of 1, the parameter β_0 tends to $-\infty$ and the four-parameter logistic is inappropri-

ate: β_0 should be constrained to a constant. If ϕ_1 is close to 0 the setting of s is probably too large, in which case it is worth halving s and repeating the optimization after recalculating the initial ϕ. If $\phi_1 = 0$ and $\phi_2 = 1$ there are insufficient points on the steepest part of the curve and the limiting form is a step function.

(7) Estimates of θ_1, θ_2, β_0, and β_1 are derived from ϕ_1 and ϕ_2 and the regression coefficients. In the limiting form one or more will be infinite, so it is preferable to redefine the model temporarily. The dispersion matrix is best computed afresh using the matrix of first derivatives with respect to θ_1 and θ_2, and must be omitted in limiting cases.

(8) Functions of parameters such as the upper asymptote, $\beta_0 + \beta_1$, may be derived, and their standard errors estimated as in Section 3.2. Other useful functions are the slopes of the curve, estimates of x given y, and predicted values of y.

(9) The determination of confidence limits for $\beta_0 + \beta_1$ is important, because the curve is often used for predicting a hypothetical upper limit to a growth process. The confidence interval is always skew, overestimates being more consistent with the data than corresponding underestimates.

(10) To test the significance of β_0 the analysis can be repeated with a fixed value of β_0, and the increase in the residual sum of squares can be compared with the residual mean square in the usual way.

Comparison of data sets depends on context. Growth curves for different organisms can be compared fitting common values of θ_1 and θ_2, allowing individual values of β_0 and β_1. Calibration trials, in which the difference between the data sets is assumed to be due solely to the difference in the origin of x, would require all data sets to have the same θ_1, β_0, and β_1 but individual values of θ_2.

5.2.2. Example of the Fitting Procedure: Mixture of Two Normal Distributions

A different type of example is that of fitting a mixture of two normal distributions by maximum likelihood estimation. The frequency function is assumed to be of the form

$$f(x; \mu_1, \mu_2, \sigma_1, \sigma_2, \alpha) = \alpha \Phi\left(\frac{x - m_1}{\sigma_1}\right) + (1 - \alpha) \Phi\left(\frac{x - m_2}{\sigma_2}\right), \quad (5.2.6)$$

where $0 < \alpha < 1$ and Φ is the standard normal frequency function. To eliminate equivalent solutions, assume also that $\mu_1 \leq \mu_2$. This problem was first studied by Pearson (1896) who proposed the solution (for $\sigma_1 = \sigma_2$) in terms of the first four moments. The method to be described is based on Ross (1970).

(1) Data may be a sample of n values of x, or a set of counts $r_1, \ldots, r_{k+1}$ corresponding to the classes $x < u_1, u_1 < x < u_2, \ldots, u_k < x$. One or more

of the values of r_i may be unknown (if the data are truncated). If $x = \log y$, the model is a mixture of two lognormals for observations y. The solution for the classified data will depend to some extent on the choice of $u_i, \ldots,$ u_k, which should take into account the method of recording x, including any rounding procedures.

(2) The log-likelihood for the n sample values takes longer to compute than the log-likelihood for $(k + 1)$ classified values, and there is a great advantage in fitting the classified data first, followed by the unclassified data if required. When truncated data are supplied, probabilities are divided by the total corresponding to the cells for which data are available.

(3) There must be at least five classes with nonzero frequencies, and no single class should include too high a proportion of the data, say 50%. Estimates of sample mean and variance are derived either from the full data or from class mid-marks, with values for the tails estimated from adjacent classes.

(4) The search for a set of five stable parameters leads to awkward equations to be solved, and so a simpler strategy is to rely on the stability of the expected mean and variance

$$m = E(x) = \alpha\mu_1 + (1 - \alpha)\,\mu_2,$$

$$v = \text{Variance}(x) = \alpha\sigma_1^2 + (1 - \alpha)\,\sigma_2^2 + \alpha(1 - \alpha)(\mu_1 - \mu_2)^2. \quad (5.2.7)$$

If m and v are assumed known, the three-parameter model estimating $\sigma_1, \sigma_2,$ and α only is easily fitted, using the equations

$$\mu_2 - \mu_1 = \sqrt{v - \frac{\alpha\sigma_1^2 - (1 - \alpha)\,\sigma_2^2}{\alpha(1 - \alpha)}},$$

$$\mu_2 = m + \alpha(\mu_2 - \mu_1), \qquad \mu_1 = \mu_2 - (\mu_2 - \mu_1), \qquad (5.2.8)$$

provided the expression under the square root is positive. The ranges of $\sigma_1, \sigma_2,$ and α are therefore constrained. There are often unwanted local optima associated with very small values of σ_1 or σ_2 in which the influence of one component is confined to a single class, and α is close to 0 or 1. These can be avoided by setting minimum values to σ_1 and σ_2. If there is only one component in reality, solutions will be found close to the subspace, $(\mu_2 - \mu_1) = 0$, and the model will be overparameterized. If there are more than two modes in the data then two or more solutions may exist, but the fit will be poor.

The full model in terms of five parameters is fitted starting from the solution for the three-parameter model, and it is usually safe to use the defining parameters of (5.2.6). A similar procedure is effective for the special case $\sigma_1 = \sigma_2$.

(5) Initial estimates of m and v are the sample mean and variance. Initial estimates, $\alpha = 0.5, \sigma_1 = \sigma_2 = 0.8\sqrt{v}$, are usually adequate, but some iterations may be saved by trying a small range of starting values and selecting the best. Even better is to fit first the special case $\sigma_1 = \sigma_2$ (which is needed in any case unless the variance are clearly different), which will generally give a good estimate of α. If the full data are available, a further optimiza-

tion is necessary but the parameters should change little. When there is truncation, with one or more missing classes, the provisional sample variance will be underestimated, but provided the proportion missing is not too great the parameters after the first stage will be adequate initial estimates for the second stage.

(6) The asymptotic dispersion matrix may be computed, but it will be unreliable except in large samples, and unobtainable if the limiting case $\alpha = 0$ or 1 is approached, or if μ_1 is close to μ_2. It is preferable to rely on comparisons with simpler models, such as $\sigma_1 = \sigma_2$, or the single normal distribution with two parameters. When truncated data are fitted the estimate of the full sample size may be computed, and by numerical differencing with respect to each parameter, estimates of its first derivatives may be used in conjunction with the dispersion matrix to estimate the variance.

The comparison of several samples requires the class limits to be the same in each sample, otherwise the analysis for the combined samples cannot be performed.

5.3. Algorithms for Nonlinear Inference

The numerical tools required for nonlinear inference using transformations of parameters are as follows.

(1) Optimization of:
   (a) the log-likelihood with respect to p parameters, θ;
   (b) the log-likelihood with respect to q out of p parameters, optimized with respect to the remaining $p - q$ parameters (separable linear parameters);
   (c) the log-likelihood with respect to q out of p parameters, the remaining $p - q$ parameters having fixed values; and
   (d) the value of a general function, subject to the constraint that the log-likelihood has a particular value.
(2) Solutions of nonlinear simultaneous equations, to compute inverse transformations.
(3) Root finding algorithms, for locating confidence intervals.
(4) Graphs and contouring algorithms, to display the functions described in Chapter 4.

5.3.1. Optimization of Log-Likelihoods

There is a large literature on optimization methods for nonlinear functions of several variables (see, e.g., Chambers, 1973). When the objective function is a log-likelihood with respect to unknown parameters there is certain extra information available which makes the problem simpler, if that information may be used.

Optimization methods are classified in several ways:

(a) the objective function is: (i) a sum of squares, (ii) a general function;
(b) the method requires: (i) computed derivatives, (ii) estimated derivatives from differencing, or (iii) function values only; and
(c) the search domain is: (i) unconstrained, (ii) constrained, to lie within or on some subspace.

The information that may be relevant if the objective function is a relative log-likelihood includes:

(a) the fact that it may be expressed as a sum of squares of deviance residuals (Section 2.4.1) which may be differentiated;
(b) that the minimum is bounded below by zero, attained when the fit is exact; and
(c) that the theory of stable parameters provides the possibility of appropriate transformations and initial estimates of transformed parameters.

5.3.2. Optimization on a Line

The basic algorithm for optimization on a line used in most practial optimization algorithms is a combination of discrete steps to bracket a minimum, followed by quadratic interpolation through three points: (x_i, y_i), $i = 1, 2, 3$, fitting the function

$$a + bx + cx^2 = \sum_{i=1}^{3} y_i \frac{(x - x_i)(x - x_k)}{(x_i - x_j)(x_i - x_k)}, \tag{5.3.1}$$

whose minimum at

$$x_m = -\frac{b}{2c}, \tag{5.3.2}$$

is

$$E(y_m) = a - \frac{b^2}{4c}, \tag{5.3.3}$$

provided $c > 0$. If the function is differentiable the minimum of the quadratic with given y' and y'' is at $-y'/y''$, and if a third derivative is available, Halley's formula based on the fitted cubic predicts a minimum at

$$\frac{-y'}{y'' - y'y'''/2y''}. \tag{5.3.4}$$

Halley's formula is not safe to use except when close to the minimum, especially when y'' is small. More useful is the method of Davidon (1969) which uses the function value and first derivative at two points to give the four parameters in the cubic. Algorithms based on Davidon's method are found in Dixon (1972), Walsh (1975), and other texts and are available, for example, in the NAG Library.

Instead of expressing the asymmetry of the function at the minimum in terms of a cubic, which is incapable of extrapolation beyond a short distance, Ross (1980) proposed that the x axis be transformed by a skewing function in such a way that four values of y should lie on a quadratic, whose minimum is then transformed back to obtain an improved estimate of the minimum.

The simplest transformation is the affine transformation

$$t = \frac{x}{1 + rx}, \qquad x = \frac{t}{1 + rt}, \tag{5.3.5}$$

with parameter r. The main disadvantage of this is the singularity at $x = -1/r$, which restricts the range of validity. The condition that four points lie on a quadratic is the vanishing of the determinant

$$\begin{Bmatrix} y_1 & 1 & t_1 & t_1^2 \\ y_2 & 1 & t_2 & t_2^2 \\ y_3 & 1 & t_3 & t_3^2 \\ y_4 & 1 & t_4 & t_4^2 \end{Bmatrix}, \tag{5.3.6}$$

or

$$(y_2 - y_1)(t_3 - t_1)(t_4 - t_1)(t_4 - t_3) + (y_3 - y_1)(t_4 - t_1)(t_2 - t_1)(t_2 - t_4)$$
$$+ (y_4 - y_1)(t_2 - t_1)(t_3 - t_1)(t_3 - t_2) = 0,$$

which may be shown to be quadratic in r, after cancelling common terms. One of the inner points may be taken as the origin, and the root of the equation closest to zero taken. The adjusted estimate of the minimum on the x scale is (to order r)

$$\frac{-b}{2c + 3br}. \tag{5.3.7}$$

The first three points are those used to estimate the quadratic. The fourth point is the estimated minimum of the quadratic, and if its value is sufficiently close to expectation the adjustment is not needed. If the predicted minimum is out of range, or too far from the other points, a nearer point may be chosen for (x_4, y_4). If the quadratic in r has no real roots the transformation is inappropriate (e.g., if there is more than one minimum).

To illustrate the effect of the transformation, consider the problem of minimizing the function

$$y = (\log x - 2)^2.$$

This is typical of nonlinear log-likelihood, a squared residual which is infinite at two points on the scale, but would be quadratic on some scale, if known. There is a point of inflexion at $\log x = 3$, and the quadratic formula is only

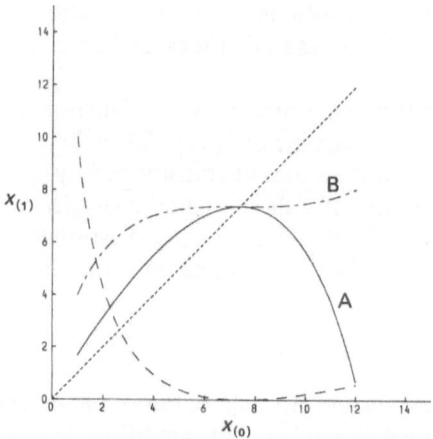

Fig. 5.1. Convergence in diagrams (see Fig. 4.28) for minimizing the function $y = (\log x - 2)^2$ by Newton's method (curve A) and by the corrected formula (5.3.7) (curve B).

valid between $x = 1$ and $x = 12$, the value of the next iteration being shown in Fig. 5.1 (curve A). Using the value at the predicted optimum, or the value at $x + 5$ or $x - 5$, if the step is greater than 5, the predicted values from (5.3.7) are shown as curve B. Estimates of r based on the discrepancy between the observed and predicted minimum are less satisfactory, unless the discrepancy is small. If $D = y_m - E(y_m)$, the assumption that y_m lies on the quadratic in the transformed scale leads to the equation

$$r^2 y'^2 (y''D - \tfrac{3}{2}y'^2) - ry'y''(2y''D + y'^2) + (y'')^3 D = 0. \qquad (5.3.8)$$

This formula is only valid if the predicted quadratic minimum is in the feasible region.

For theoretical studies it is possible to use the condition that the third derivative on the transformed scale is zero. This can be shown to lead to the equation

$$6y'r^2 - 6y''r + y''' = 0, \qquad (5.3.9)$$

which gives very similar estimates of r when the trial point is close to the optimum of equations (5.3.6) and (5.3.8). The smallest value of r is then used to adjust the estimated minimum using equation (5.3.7).

5.4. Separable Linear Parameters

Separable models have been introduced in earlier chapters (Sections 2.1.4, 3.1.2, and 4.4.3.). The following discussion relates to computing the solutions.

The fact the linear parameters in models may be estimated directly as functions of nonlinear parameters is surprisingly little exploited, even though it was used long before the electronic computer was invented. Fisher (1939) was certainly familiar with the concept of curve fitting using trial values of nonlinear parameters, and the textbooks of Snedecor (1937) and Goulden (1939) suggest graphical methods for improving on a fixed choice of parameter values.

Hartley (1968) used the term "internal least squares" to describe a method for fitting nonlinear curves which could be expressed as solutions of differential equations, the simplest example being the exponential curve; the data had to be equally spaced. Stevens (1951) showed that the Gauss–Newton equations for the parameters of the exponential curve could be expressed in terms of the nonlinear parameters only, a method that was rediscovered by Barham and Drane (1972). Richards (1961) adopted a direct numerical approach, in which the residual sum of squares is minimized with respect to the nonlinear parameters, and showed how to reconstruct the dispersion matrix for the complete set of parameters. Ross (1970) described how the direct method could be incorporated into a general optimization program, and how it could be further improved by choice of stable nonlinear parameters. Lawton and Sylvestre (1971) advocated a similar approach, but appear to have applied it also to the Gauss–Newton algorithm in a two-stage process, described formally by Ross (1982).

The most important cases where linear parameters are separably estimable are as follows:

(1) *Normal errors*, $E(y) = \sum \beta_j f_j(\theta)$, where $f_i(\theta)$ are constants or functions involving nonlinear parameters θ. Examples are:

$$E(y) = \beta_0 - \beta_1 \theta_1^x,$$

$$E(y) = \beta_0 + \beta_1 \theta_1^x + \beta_2 x,$$

$$E(t) = \beta_0 + \frac{\beta_1}{1 + \exp(-\theta_1(x - \theta_2))},$$

$$E(y) = \frac{\beta_1 x}{x + \theta_1}.$$

Given θ, the estimates $\beta(\theta)$ are the standard weighted linear regression coefficients of y on the variates f_j.

(2) *Poisson variables* with expectation $\beta f(\theta), f > 0$. The log-likelihood,

$$\sum y_i \log(\beta f_i) - \beta \sum f_i, \tag{5.4.1}$$

maximized with respect to β, gives the estimate

$$\hat{\beta}(\theta) = \frac{\sum y_i}{\sum f_i}. \tag{5.4.2}$$

A practical example is known as Wadley's problem in bioassay (Finney, 1978) in which y_i is the number of survivors in a sample of unknown size, β, subjected to treatment x_i, where $f(x_i, \theta)$ is the theoretical proportion of survivors.

Another important application is to the fitting of truncated frequency distributions where observations outside a given range are not recorded. The parameter β is then the total sample size that would have been observed, including the truncated observations.

(3) *Gamma variables* with expectation $\beta f(\theta), f > 0, y > 0$. The log-likelihood

$$\sum \left(\frac{y_i}{\beta f_i} \right) + \log(\beta f_i), \tag{5.4.3}$$

maximized with respect to β gives the estimate

$$\hat{\beta}(\theta) = \frac{1}{n} \sum \left(\frac{y_i}{f_i} \right). \tag{5.4.4}$$

This model is likely to occur whenever the gamma distribution is used to model continuous positive variables, such as reaction times. A scale parameter β is nearly always required.

The three main algorithms using separable linear parameters are therefore:

(1) Direct minimization of the reduced log-likelihood with respect to p nonlinear parameters (Richards, 1961).
(2) Gauss–Newton type minimization of sums of squares with respect to $p + 2$ parameters, and computing adjustments to the nonlinear parameters (Stevens, 1951; Barham and Drane, 1972).
(3) Gauss–Newton type minimization of deviance residuals, substituting fitted linear parameters (Lawton and Sylvestre, 1971; Ross, 1970).

To illustrate the differences between the three approaches, consider the separable model

$$E(y) = \beta_0 + \beta_1 \theta^x$$

fitted by least squares to the following data:

x	0	1	2	3
y	3	12	17	20

starting with $\theta = 0.4$ and $\delta\theta = 0.01$. The optimum θ is 0.57026.

METHOD 1. The residual sum of squares for linear regression on θ^x, $R(\theta)$, is

$$R(0.39) = 2.819228, \qquad R(0.40) = 2.498507, \qquad R(0.41) = 2.198882,$$

and the predicted minimum of the quadratic through these values is at $\theta = 0.54703$.

METHOD 2. The adjustment to r is computed from the regression equation

$$y = \beta_0 + \beta_1 \theta^x + (\beta_1 d\theta) x\theta^{x-1},$$

where

$$\theta^* = (1, 0.4, 0.16, 0.064),$$

and

$$x\theta^{x-1} = (0, 1, 0.8, 0.48),$$

yielding the equations

$$
\begin{pmatrix}
4 & 1.624 & 2.28 \\
1.624 & 1.1897 & 0.5587 \\
2.28 & 0.087 & 1.8704
\end{pmatrix}
\begin{pmatrix}
\beta_0 \\
\beta_1 \\
\beta_1 d\theta
\end{pmatrix}
=
\begin{pmatrix}
52 \\
11.8 \\
35.2
\end{pmatrix},
$$

whence $\beta_1 = -19.4853$, $\beta_1 d\theta = -2.7875$ and $d\theta = 0.14306$ and $\theta = 0.54306$.

METHOD 3. The residuals after fitting $\theta = 0.4$ and $\theta = 0.41$ are as follows:

$$e(0.4) = 0.4295, -1.1053, -0.3193, 0.9951,$$

$$e(0.41) = 0.4110, -1.0311, -0.3123, 0.9324,$$

whence

$$\frac{\partial e}{\partial \theta} = -1.85, 7.42, 0.70, -6.27,$$

and

$$d\theta = -\sum e \frac{\partial e}{\partial \theta} \Big/ \sum \left(\frac{\partial e}{\partial \theta} \right)^2 = 0.15729,$$

whence $\theta = 0.55729$.

The three methods will now be discussed in detail.

5.4.1. Direct Minimization of Reduced Log-Likelihoods

The reduced log-likelihood, $R(\theta)$, computed for normal error models by weighted or unweighted linear regression (with on without an intercept term), and for Poisson and gamma models by substituting $\hat{\beta}(\theta)$ using formulas (5.4.1)–(5.4.4), may be minimized using any of a number of techniques for general optimization. The estimated linear parameters $\hat{\beta}(\hat{\theta})$ are the values obtained at the optimum $\hat{\theta}$. The estimated dispersion matrix may be obtained either directly, by expressing the model in terms of all the θ and β parameters simultaneously or indirectly, following Richards (1961), who showed that given $\hat{\beta}(\hat{\theta})$ and the matrices

$$\underset{p \times p}{V} = \left(\frac{\partial^2 R(\theta)}{\partial \theta_i \, \partial \theta_j} \right)^{-1}_{\theta = \hat{\theta}}, \qquad (5.4.5)$$

$$\underset{p \times p}{A} = \left(\frac{\partial \hat{\beta}_i}{\partial \theta_i} \right), \qquad (5.4.6)$$

$$B_{q \times q} = \left(\frac{\partial^2 L}{\partial \beta_i \, \partial \beta_j} \right)^{-1}_{\theta = \hat{\theta}}, \tag{5.4.7}$$

then the full dispersion matrix of $(\hat{\theta}, \hat{\beta})$ is

$$\begin{pmatrix} V & VA \\ A'V & B + A'VA \end{pmatrix}. \tag{5.4.8}$$

The matrix V is output by the optimization algorithm and the matrix β by the regression algorithm (or by computing the single term $b_{11} = \hat{\beta}^2 / \sum y$ for the Poisson model and $b_{11} = \hat{\beta}/n$ for the gamma model). The matrix A must be obtained by differencing, either centrally,

$$\frac{\partial \hat{\beta}}{\partial \theta} = \frac{\hat{\beta}(\theta + d\theta) - \hat{\beta}(\theta - d\theta)}{2d\theta},$$

or noncentrally

$$\frac{\partial \hat{\beta}}{\partial \theta} = \frac{\hat{\beta}(\theta + d\theta) - \hat{\beta}(\theta)}{d\theta}.$$

Central differences are more reliable, but if the curvature of $\hat{\beta}$ is ignored noncentral differences may be used with little loss of accuracy.

As a simple example, consider the model

$$E(y) = \beta \theta^x,$$

where y has the Poisson distribution, fitted to the following data:

x	1	2	3	4	5
y	19	13	14	11	6

The optimum value of θ is close to 0.8, and we find the following values for the reduced log-likelihood and for $\hat{\beta}$:

θ.	$R(\theta)$	$\hat{\beta}(\theta)$
0.796	0.594857	23.729
0.797	0.594801	23.053
0.798	0.594930	23.577

whence

$$\hat{\theta} = 0.7968, \qquad \hat{\beta} = 23.668,$$

$$V_{11} = 1/185 = 0.005405,$$

$$a_{11} = \frac{23.577 - 23.729}{0.002} = -76,$$

$$b_{11} = \frac{\hat{\beta}^2}{\sum y} = \frac{(23.668)^2}{66} = 8.4875,$$

and

$$D^2(\hat{\theta}, \hat{\beta}) = \begin{pmatrix} 0.005405 & -0.41081 \\ -0.41081 & 39.7091 \end{pmatrix}.$$

Direct evaluation of the inverse Hessian of the log-likelihood with respect to θ and β gives the matrix

$$\begin{pmatrix} 0.005377 & -0.40807 \\ -0.40807 & 39.8527 \end{pmatrix}.$$

The matrix (5.3.8) is very important. It shows that:

(i) if we are not interested in the variances of the β parameters then the dispersion matrix of the θ parameters is simply V;

(ii) if B is very small the matrix is nearly singular;

(iii) The dispersion matrix of β has two components, B due to uncertainty in $\hat{\beta}$ given $\hat{\theta}$, and $A'VA$ due to uncertainty in $\hat{\theta}$.

5.4.2. Transformations To Improve Direct Minimization

The difficulties experienced when minimizing $R(\theta)$ include the following:

(i) *Finite differences*
Explicit formulas for derivatives are too complicated in most cases, and so finite differences must be used. $R(\theta)$ must therefore be computed to high precision, and step lengths must be sufficiently large to avoid distortions due to rounding or truncation error.

(ii) *Singularities*
For some values of θ the equations for $\hat{\beta}$ are singular or nearly so, leading to irregularities in the estimated $R(\theta)$. The singularity may be avoided in particular known cases by rewriting the functional form, which will ususally affect the β estimates at the same time. A well-known example is exponential regression,

$$y = \beta_0 + \beta_1 \exp(-\theta x),$$

where if $\theta = 0$, $f(\theta) = 1$ for all x and β_0 and β_1 are infinite. If, instead, we write

$$y = \beta_0' + \frac{\beta_1'[\exp(-\theta x) - 1]}{\theta}, \qquad \theta \neq 0,$$

$$= \beta_0' + \beta_1' x, \qquad\qquad\qquad \theta = 0,$$

$f(\theta)$ is continuous at $\theta = 0$, and the β values are those for linear regression.

A similar example is provided by the rectangular hyperbola,

$$y = \beta_0 + \frac{\beta_1}{1 + \theta x},$$

which can be rewritten

$$y = \beta_0' + \frac{\beta_1' x}{1 + \theta x}$$

to avoid the singularity at $\theta = 0$.

In general, if θ_s is a singular value of θ which makes $f(\theta)$ constant, it is better to use

$$\frac{f(\theta) - f(\theta_s)}{f'(\theta_s)}, \qquad (5.4.9)$$

which is a linear function of $f(\theta)$ and therefore gives the same fit to the data.

An alternative device is to approximate $f(\theta)$ by a polynomial in the close neighborhood of θ_s, or to set parameter limits that enable one to avoid the region close to θ_s. For example, the exponential curve is approximated by writing

$$f(\theta) = x\left(1 - \frac{\theta x}{2}\right)$$

when θ is sufficiently small.

(iii) *Upper bounds to $R(\theta)$*

In the normal error model, $R(\theta)$ is bounded above by the sum of squares obtained when the $f(\theta)$ are uncorrelated with y. At the point θ_u where this occurs there is a local maximum, $R_{\max}$, and if θ_u is not at infinity $R(\theta)$ cannot be realistically approximated by a quadratic function, and in the neighborhood of θ_u, $R(\theta)$ would be convex rather than concave, providing poor information on the location of minima.

When

$$E(y_i) = \beta_0 + \beta_1 f_i(\theta), \qquad R_{\max} = \sum (y_i - \bar{y})^2,$$

and when

$$E(y_i) = \beta_1 f_i(\theta), \qquad R_{\max} = \sum y_i^2,$$

and when

$$E(y_i) = \beta_0 + \beta_1 x + \beta_2 f_i(\theta),$$

$R_{\max}$ is the linear regression residual sum of squares. Ross (1980) proposed that the function

$$S(\theta) = \frac{R(\theta)}{R_{\max} - R(\theta)} \qquad (5.4.10)$$

should be optimized, rather than $R(\theta)$, with appropriate safeguards when $R(\theta)$ is very close to $R_{\max}$. The optima of $R(\theta)$ and $S(\theta)$ are identical, but $S(\theta)$ is

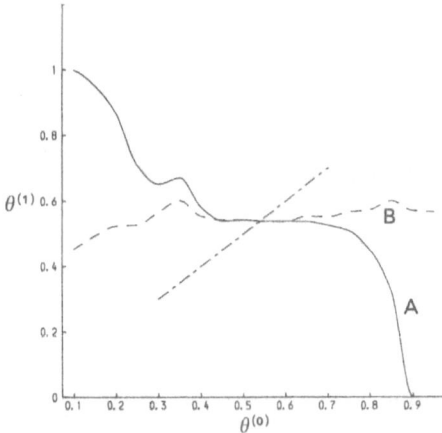

Fig. 5.2. Convergence diagrams (see Fig. 4.28) for the nonlinear parameter θ in the simple exponential model using the residual sum of squares (curve A) and the transformed function $S(\theta)$ (curve B).

convex over a wider region and is less liable to diverge. Near the optimum the two functions are very similar, becasue $R_{max} - R(\theta)$ is nearly constant.

For a line search, the transformation from R to S decreases the Newton step predicted by fitting a quadratic, as can be seen from the following equation:

$$\delta\theta = \frac{-dS}{d\theta} \Big/ \frac{d^2S}{d\theta^2}$$

$$= \frac{-dR}{d\theta} \Big/ \left(\frac{d^2R}{d\theta^2} + \frac{2}{R_{max} - R} \left(\frac{dR}{d\theta} \right)^2 \right), \qquad (5.4.11)$$

which shows that in relative terms the second derivative is increased. The form of the Hessian matrix of second derivatives of S rather than of R tends to reduce ill-conditioning, and widens the zone of convergence. Near the solution the terms in $dR/d\theta$ tend to zero and the convergence behavior of the objective functions coincides. The need for stable parameters is of course unaffected.

An example of the effect of the transformation is illustrated in Fig. 5.2.

5.4.3. Stevens' Method

The method described by Stevens (1951) to fit the exponential regression model

$$E(y) = \beta_0 + \beta_1 \theta^x$$

applies to some, but not all models of the form

$$E(y) = \sum \beta_j f_j(\theta, x). \qquad (5.4.12)$$

Expanding $f_j(\theta)$ about θ_0 in Taylor series, writing $\delta\theta = \theta - \theta_0$,

$$E(y) = \sum_{j=1}^{q} \beta_j \left(f_j(\theta) + \sum_k \frac{\partial f_j}{\partial \theta_k} \cdot \delta\theta + O(\delta\theta' \cdot \delta\theta) \right),$$

the model becomes a linear regression model of y on q variables $f_j(\theta)$ with coefficients β_j, and on up to pq variables $\partial f_j/\partial \theta_k$ with coefficients $\beta_j \delta\theta_k$. Stevens' method applies to those special cases in which each nonlinear parameter occurs in only one expression $f_j(\theta)$, so that all $\delta\theta_k$ adjustments may be computed uniquely at each iteration, by dividing out the estimate of β_j.

The method is closely related to iteratively reweighted least squares (Green, 1984), in that it relies entirely on linear regression calculations without modification if the residual sum of squares is not reduced. It therefore has a limited convergence zone and requires good initial estimates of the nonlinear parameters.

5.4.4. Two-Stage Gauss–Newton Method

The simple Gauss–Newton method of minimizing a sum of squares $\sum e_1^2(\theta)$ is to expand each $e_i(\theta)$ in Taylor series, and taking the linear term only, to write

$$e_i(\theta) = e_i(\theta_0) + \sum \frac{\partial e_i}{\partial \theta_k} \delta\theta_k,$$

or

$$e_i(\theta_0) = -\sum \frac{\partial e_i}{\sigma \theta_k} \delta\theta_k + e_i(\theta), \tag{5.4.13}$$

which is in the form of a linear regression of $e_i(\theta_0)$ on p variables $-\partial e_i/\partial \theta_k$, $k = 1, \ldots, p$.

The solution is therefore

$$\delta\theta_k = -\left(\frac{\partial e'}{\partial \theta_k} \cdot \frac{\partial e'}{\partial \theta_k} \right)^{-1} \left(\frac{\partial e'}{\partial \theta_k} \cdot e \right), \tag{5.4.14}$$

which is the Gauss–Newton step. Geometrically, the Gauss–Newton point corresponds to the foot of the perpendicular from the data point to the tangent at θ_0 to the solution locus (4.3.1), and the parameter value θ_1 is determined from the linear parameter system whose gradient is the value of $\partial e/\partial \theta$ at θ_0. The value, θ, then determines a new point on the solution locus.

The *two-stage Gauss–Newton* process is based on the same geometrical idea except that the solution locus is reduced to q dimensions by substituting the $(p - q)$ fitted linear parameters $\hat{\beta}(\theta)$, and operating on the nonlinear parameters only. Numerical diffferences are used for $\partial e/\partial \theta$ which is rarely of simple differentiable form. The algebra is tedious rather than impracticable. For example, to fit the model

$$E(y_i) = \beta_0 + \beta_1 f_i(\theta)$$

for a single nonlinear parameter θ we compute

$$S_{yf} = \sum (y_i - \bar{y})(f_i - \bar{f}), \qquad S_{ff} = \sum (f_i - \bar{f})^2,$$

$$e_i = y_i - \bar{y} - \frac{f_i S_{yf}}{S_{ff}},$$

when

$$\frac{\partial e_i}{\partial \theta} = \frac{-\partial f_i}{\partial \theta} \frac{S_{yf}}{S_{ff}} - \frac{f_i}{S_{ff}^2}\left(S_{ff}\frac{\partial}{\partial \theta}(S_{yf}) - S_{yf}\frac{\partial}{\partial \theta}(S_{ff})\right),$$

and

$$\delta\theta = -\sum e_i \frac{\partial e_i}{\partial \theta} \bigg/ \sum \left(\frac{\partial e_i}{\partial \theta}\right)^2$$

These formulas may be used to study convergence in special cases, but in general it is simpler and more reliable to pass the functions e_i to a direct algorithm for minimizing a sum of squares without computing derivatives.

5.4.5. Numerical Comparison of Four Methods

The methods discussed above may be compared for the exponential example (5.3.3) using convergence loci (Section 4.4.8) to show the relationship between $\theta_{(0)}$ and $\theta_{(1)}$ for a single iteration. The flatter the curve as it crosses the line $\theta_{(0)} = \theta_{(1)}$, the faster the rate of convergence. In Fig. 5.3 there are four curves:

A: Direct Newton method on $R(\theta)$, diverging if $\theta_{(0)} > 0.94$.
B: Newton method on $S(\theta)$, converging for all $\theta_{(0)} < 1$.
C: Stevens' method, diverging if $\theta_{(0)} > 0.77$.
D: Two-pass Gauss–Newton method, converging for all $\theta_{(0)} < 1$.

Method D is clearly superior for a wide range of initial estimates, whereas methods A and C require some care in the choice of initial estimate, as was recognized by Stevens (1951). A simple change of scale of x will affect all the curves in Fig. 5.3, but there is no practical way of knowing *a priori* which scale is best for a particular method.

While the differences between the methods may not be too great in one dimension, for many-parameter problems the differences are more important. For the nonlinear parameters it is possible to construct convergence contours, as described in Section 4.4.8. The methods should be illustrated on a stable parametrization, otherwise the comparisons are not fair because in a practical algorithm there could be checks to prevent divergence. The example chosen here is a logistic curve with a single linear parameter, fitted to an artificial data set. Two sets of contours are plotted in Fig. 5.4: the solid lines represent points from which the Newton method applied to $S(\theta)$ reduce the function to 0.00270 in one and two iterations, respectively, while the dotted lines represent points from which the two-pass Gauss–Newton method gives the equivalent sum of squares in one and two iterations. The direct Newton method on $R(\theta)$

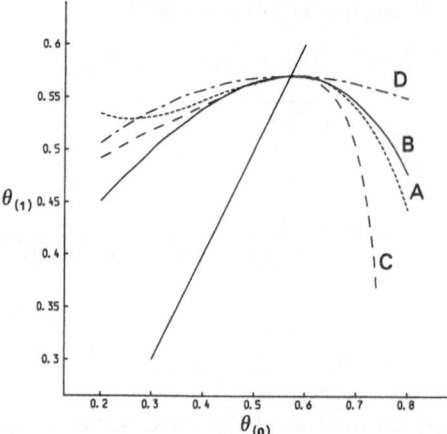

Fig. 5.3. Convergence diagrams (see Fig. 4.28) for the nonlinear parameter in the exponential model using four different algorithms (see text).

gives almost identical results for the inner contour but a slightly smaller region for the outer contour. Stevens' method gives similar results to the two-pass Gauss–Newton, but diverges rapidly for points outside the outer contour, because trial values outside the triangle, $0 < \theta_1 < \theta_2 < 1$, give unacceptable shapes of curve. It would be possible to modify Stevens' method in the same way as the other methods by combining it with a line search if the sum of squares is not reduced. Note that one method is not uniformly superior to the others; that would imply that one contour is completely contained within the others.

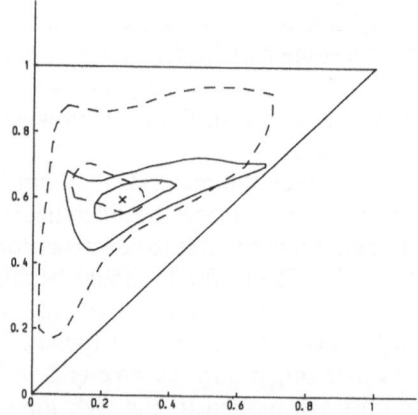

Fig. 5.4. Convergence contours (see Fig. 4.29) for two nonlinear parameters in a logistic model, using two different algorithms: ———, Newton's method applied to residual sum of squares; – – – – – – – –, two-stage Gauss–Newton method (see text).

5.5. Confidence Intervals for Parameters and Functions

Likelihood-based confidence regions for θ were discussed in Sections 1.2.2 and 3.2.1 and their plots were discussed in Section 4.4.1. Here we discuss the numerical algorithms for finding explicitly the maximum and minimum of $g(\theta)$ subject to the constraint $L(\theta) = L_c$, where $g(\theta)$ is a general continuous function of θ.

We assume that $\hat{\theta}$ has been obtained, and its approximate dispersion matrix V, so that the log-likelihood has the form

$$L(\theta) = L(\hat{\theta}) + \tfrac{1}{2}(\theta - \hat{\theta})'V^{-1}(\theta - \hat{\theta}) + O(\theta - \hat{\theta})^3;$$

as shown by Ross (1978) the standard Lagrange technique may be applied to the function

$$g(\theta) - \lambda(L(\theta) - L_c),\tag{5.5.1}$$

to obtain the equations

$$\frac{\partial g}{\partial \theta_k} = \lambda \frac{\partial L}{\partial \theta_k} = \lambda V^{-1}(\theta_k - \hat{\theta}_k) + O(\theta - \hat{\theta})^3, \qquad k = 1, \dots, p,$$

$$L(\theta) = L_c.$$

Ignoring higher-order terms and substituting

$$(\theta_k - \hat{\theta}_k) = \frac{1}{\lambda} V \frac{\partial g}{\partial \theta_k}\tag{5.5.2}$$

in the expression for $L(\theta)$ we obtain

$$\lambda^2 = \left(\frac{\partial g}{\partial \theta}\right)' V \frac{\partial g}{\partial \theta} \Big/ 2(L_c - L(\hat{\theta})).\tag{5.5.3}$$

Taking λ with appropriate sign, the equations may be solved for θ if $\partial g/\partial \theta_k$ is constant, the two roots corresponding to maxima and minima of the function g.

For a linear function of a linear normal model, the solution is direct and noniterative. If $g(\theta) = a'\theta$,

$$\theta = \hat{\theta} \pm \sqrt{\frac{2(L_c - L(\theta))}{a'Va}}\, Va.\tag{5.5.4}$$

For the special case in which $g(\theta)$ is one of the parameters itself, θ_k,

$$\theta = \hat{\theta} \pm \sqrt{2(L_c - L(\theta))}\,\frac{v_k}{\sqrt{v_{kk}}},\tag{5.5.5}$$

where v_k is the kth column vector of V and v_{kk} is the kth diagonal element. These θ are the points of contact of tangent primes, $\theta_k = $ constant, to the critical contour.

For general log-likelihoods and general functions, exact algorithms can be described, subject to the following conditions:

(1) The critical contour L_c is simple, closed, and convex in the sense that any straight line from θ outwards crosses L_c once and once only.
(2) The function $g(\theta)$ is continuous and monotone within the contour L_c.
(3) The family of surfaces, $g(\theta)$ constant, intersect with the critical contour in a single, closed domain, or not at all, so that there are no double tangents or local optima.

In practice we cannot ensure that these conditions hold, but sensible use of stable parameter transformations will support condition (1), and transformations of g may be found to support condition (2). For example, if $g(\theta) = \theta_2/\theta_1$ and θ_1 may be zero while θ_2 is always positive, the reciprocal function θ_1/θ_2 should be used instead. Condition (3) cannot be transformed away, and examples can easily be constructed of functions with discontinuous confidence intervals; for example, the quadratic polynomial is a linear model, but the confidence interval for the value of x given y may consist of two separate ranges, one for each root.

5.5.1. Exact Confidence Limits for Parameters

The exact procedure when $g(\theta) = \theta_k$ is as follows:

(1) Starting with the expected position of the upper or lower point of contact of the tangent prime given by (5.4.5), check that the point θ is feasible. If not, move inwards towards $\hat{\theta}$ to the nearest feasible value on the line joining θ to $\hat{\theta}$.
(2) Optimize $L(\theta)$ with respect to all parameters except θ_k, to obtain a minimum value $M(\theta_k) = L_c$. For a linear model $M(\theta_k) - L(\hat{\theta})$ is proportional to $(\theta_k - \hat{\theta})^2$, hence the function

$$N(\theta_k) = \sqrt{N(\theta_k)} - L(\hat{\theta}) \qquad (5.5.6)$$

should be used for linear interpretation to solve

$$N(\theta_k)' = \sqrt{L_c - L(\theta)} = N_c. \qquad (5.5.7)$$

(3) Find $N(\theta_k + d\theta_k)$ by optimization subject to $\theta_k + d\theta_k = $ constant. Then

$$\theta_k^{(1)} = \theta_k + d\theta_k\left(\frac{N_c - N(\theta_k)}{N(\theta_k + d\theta_k) - N(\theta_k)}\right).$$

If $\theta_k^{(1)}$ is further from θ than θ_k, test that it is not too far. If $\theta_k^{(1)}$ moves inwards towards $\hat{\theta}$, test that it does not pass $\hat{\theta}$.
(4) Find $N(\theta_k^{(1)})$ and solve (5.4.7) by a secant algorithm, using the two values such that $N(\theta)$ is closest to N_c at each stage, until acceptable convergence.

For each optimization step some time may be saved by choosing initial values of the active parameters on the line joining the previous solution to $\hat{\theta}$, on the grounds that the solutions within each subspace for θ_k tend to lie on a straight line. When the change in θ_k is small it will hardly be worth the trouble to predict the new values, and the current values should be adequate starting values.

If the contour L_c is not closed the solution may not exist, and a limit should be placed on the number of iterations, both in each subsidiary optimization and in the secant process. The algorithm is of course likely to take considerably more time than the original optimization of θ, typically $10p$ times, where p is the number of parameters.

5.5.2. Approximate Limits for a Function of Parameters

As in formula (5.5.4) the expected values of θ for which $g(\theta)$ will be a maximum or minimum are

$$\theta = \hat{\theta} \pm \sqrt{\frac{2(L_c - L(\theta))}{\left(\frac{\partial g}{\partial \theta}\right)' V \left(\frac{\partial g}{\partial \theta}\right)}} \, V \frac{\partial g}{\partial \theta}. \tag{5.5.8}$$

Estimates of $\partial g/\partial \theta$ are obatined either analytically or by differencing, and initially the estimates apply to $\theta = \hat{\theta}$. For approximate purposes (5.5.8) may be used noniteratively. However, in general, θ is not on the critical contour, and $\partial g/\partial \theta$ has changed in between $\hat{\theta}$ and θ. This suggests the following algorithm:

(1) From the initial value in (5.5.8) check that $L(\theta) = L_c$. If not, use the secant algorithm of the previous section to find the point $\theta^{(1)}$ or the line through θ and $\hat{\theta}$ for which $L(\theta) = L_c$.
(2) Evaluate $\partial g/\partial \theta$ for the new $\theta^{(1)}$. Substitute in (5.5.8) to obtain a new direction of search, and repeat step (1).

In practice, the algorithm does not always converge rapidly, because the change in $\partial g/\partial \theta$ may considerably affect the direction of of search. It is therefore worthwhile incorporating Aitken's acceleration technique, using three successive estimates, $\theta^{(1)}$, $\theta^{(2)}$, and $\theta^{(3)}$,

$$\theta_j = \theta_j^{(3)} - \frac{(\theta_j^{(3)} - \theta_j^{(2)})^2}{\theta_j^{(3)} + \theta_j^{(1)} - 2\theta_j^{(2)}} \tag{5.5.9}$$

applied only when the sequence is oscillating slowly.

The converged value θ^* then gives the corresponding $g(\theta^*)$. This value is not the true solution, because that requires (5.4.2),

$$\frac{\partial g}{\partial \theta_k} = \lambda \frac{\partial L}{\partial \theta_k},$$

which means that the gradient of the function is parallel to the gradient of $L(\theta)$ at $\cdot\theta^*$ rather than at the ideal point on the ellipsoidal approximation, $(\theta - \hat{\theta})'V^{-1}(\theta - \hat{\theta})$. The accuracy of the method depends on the extent to which this matters. If θ is a fairly stable parameter system the estimate of θ^* may be in error but $g(\theta^*)$ may be very close to the curve at value. The true value may be found by applying algorithm (5.5.1) with $g(\theta)$ as a parameter. The advantage of the Lagrange algorithm described here is in speed of solution and ease of computation.

5.5.3. Approximate Limits for Parameters Using Stable Transformations

When a stable parametrization $\phi(\theta)$ has been found which is assumed to have an exact quadratic representation of the critical contour, $L(\phi) = L_c$, it is possible to compute approximate limits for θ using formula (5.5.8), assuming that the transformation $\theta(\phi)$ is easily computed.

When $\phi(\theta)$ is available but $\theta(\phi)$ is not, the following procedure may be used:

(1) Given $\hat{\phi}$, V_ϕ, and $\phi(\theta)$, invert the square matrix $(\partial\phi/\partial\theta)$ to obtain $(\partial\theta/\partial\phi)$. Then the limits for θ_k are at

$$\phi^* = \phi \pm \sqrt{\frac{2(L_c - L(\phi))}{\left(\dfrac{\partial\theta}{\partial\phi}\right)'V_\phi\left(\dfrac{\partial\theta}{\partial\phi}\right)}}\, V_\phi\left(\frac{\partial\theta_k}{\partial\phi}\right). \qquad (5.5.10)$$

To solve $\phi(\theta) = \phi^*$ for θ, start with $\theta^{(0)} = \hat{\theta} + (\partial\theta/\partial\phi)(\phi^* - \hat{\phi})$, evaluate $(\partial\theta/\partial\phi)\,\theta^{(0)}$, and iterate as follows:

$$\theta^{(1)} = \theta^{(0)} + \frac{\partial\theta}{\partial\phi}(\phi^* - \phi(\theta^{(0)})),$$

which should converge fairly rapidly. This algorithm is quick because it involves only the solution of $\theta(\phi^*)$. Its accuracy depends on the adequacy of the quadratic approximation.

An important application of this algorithm is to the parameters of fitted curves. In Section 2.2.3 it was shown that when fitting a curve $y = f(x)$ with p parameters it is usually possible to find p values of x for which the fitted values of y are nearly uncorrelated; the sum of squares of the correlations are minimized using an optimization algorithm.

Let ϕ be the set of p uncorrelated ordinates, $j = 1, \ldots, p$, which may be treated as a set of stable parameters. Their variances are calculated from the dispersion matrix of θ and their covariances are assumed to be zero, and the log-likelihood is assumed to be quadratic because fitted values within the data range are approximately normally distributed. The derivatives $\partial\phi/\partial\theta$ are simple functions of x and θ, and so the required quantities $\partial\theta/\partial\phi$ may be obtained by inverting the $p \times p$ matrix $(\partial\phi/\partial\theta)$. Equation (5.5.9) is particularly simple because V_ϕ is assumed and so for θ_j,

$$\phi_1^* = \phi_i \pm \sqrt{\frac{pF\alpha(p, n - p)}{\sum_k \left(\dfrac{\partial \theta_j}{\partial \phi_k}\right)^2 \sigma^2(\phi_k)}} \left(\frac{\partial \theta_j}{\partial \phi_i}\right) \sigma^2(\phi_i), \tag{5.5.11}$$

and it is only necessary to find the values of θ corresponding to ϕ_1^* (which is only possible if ϕ^* corresponds to feasible θ).

Note that in this algorithm it is not checked that $L(\theta) = L_c$, and hence there may be some loss of accuracy at the expense of speed. The algorithm may of course be further developed to give limits for functions of θ, since $\partial g/\partial \phi_i = (\partial g/\partial \phi_i)$ may be substituted for $\partial \theta_j/\partial \phi_i$ in (5.5.10) and the required result is $g(\theta(\phi^*))$. See Example 3.12, Section 3.2.2.

CHAPTER 6

Practical Applications of Nonlinear Modeling

6.1. Some Questions To Be Asked in any Practical Application

In this chapter, I shall discuss various applications of nonlinear models; how models have been chosen and whether alternative models are an improvement.

The following questions should always be asked in any practical application:

(a) *Purpose of model fitting*
What is the purpose of fitting a model? Do we wish to predict from the model, or to estimate certain parameters, or to find a satisfactory explanation for the data, or simply to show that data exist which fit the model?

(b) *Choice of model*
Is the model chosen because of some underlying theory, or becasue it is expected to resemble the data, or because it is the conventional model used by others, or because it is algebraically and statistically tractable?

(c) *Alternative models*
What alternative models might be chosen, and what difference does it make to the purpose of fitting? What special cases and generalizations are most appropriate?

(d) *Error distributions*
Are the observations independently and normally distributed with equal variance? And if not, what difference does it make? If the observations must

be positive, or less than a maximum value, how should observations near the limits be weighted?

(e) *Design*
Is the distribution of controllable variables (times of observation, replication, etc.) adequate for the purpose of fitting? What major alterations would one recommend for future invesitgations?

(f) *Applicability of results*
To what extent may the fitted model be applied in different circumstances? Which parameters are specific to the given data and which may be considered of wider application?

6.2. Curve Fitting to Regular Observations in Time

Numerous applications of curve fitting use regular published statistics of human activity. The growth of populations, enterprises, or sales of goods is plotted against time, and if a curve is suggested the fit is used for prediction. The main justification for this practice is that the fit is good. If the time series were stationary, or upset by major economic crises or wars or natural disasters, there would be no question of trying to fit a curve. Yet major policy decisions depend on the forecasts, which may require action many years later.

Time series models (as described, e.g., by Box and Jenkins, 1970), will not be discussed here. There is much work to be done to elucidate the relationships between the parameters of the underlying trend and the autoregressive and moving-average parameters in a Box–Jenkins model.

In previous chapters where growth curves have been discussed (e.g., Section 2.2.1) it has been assumed that the observations are in some sense repeatable. Studies of the growth of plants, animals, and biological communities can be replicated, and the parameters common to all replicates assumed to apply to further examples of the same class. Environmental changes, or incidents such as disease in animals, require the particular analysis to be disregarded, on the grounds that the curve may only be fitted properly in the absence of such effects. Defects in design may be remedied by altering the frequency of measurement and extending the period of observation. Such considerations are inapplicable when the data are unique and historical, and predictions can apply only to the future.

6.2.1. Electricity Consumption in Great Britain, 1950–1985

In Table 6.1 the total consumption of electricity in Great Britain, as published in *Annual Abstracts of Statistics*, is shown. In Fig. 6.1 the figures are plotted

Table 6.1. Electricity consumption in Great Britain (terawatt-hours per year).

Year	Consumption	Year	Consumption	Year	Consumption
1950	45.5	1962	122.4	1974	213.9
1951	50.5	1963	133.9	1975	213.5
1952	52.0	1964	140.4	1976	216.3
1953	55.6	1965	151.1	1977	220.9
1954	61.4	1966	156.9	1978	225.3
1955	67.4	1967	161.7	1979	235.3
1956	73.5	1968	173.9	1980	225.1
1957	77.2	1969	185.4	1981	222.0
1958	83.9	1970	193.9	1982	217.1
1959	90.5	1971	199.4	1983	230.4
1960	102.4	1972	205.4	1984	224.7
1961	110.2	1973	220.6	1985	236.0

SOURCE: *Animal Abstracts of Statistics*, HMSO.

on a logarithmic scale. Up to 1965 the linearity of the plot was so striking that it became axiomatic that growth was exponential. Advertisements in the late 1960s proclaimed that "the demand for electricity doubles every ten years." Yet by 1970 the slope was declining, and after the energy crises of 1973–74 the plot resembles a stationary times series. The forecast for 1985 using linear regression on the log scale on the data up to 1965 is 785 terawatt-hours (TWh), over 3.3 times the actual consumption, yet the standard error of prediction (in conventional regression terms) is only 27.1 TWh. What went wrong?

The counsel of despair is to forbid the use of extrapolated curves under any circumstances. The favorite excuse is that the changed environment could not have been anticipated. A more imaginative approach is to examine the implications of continued growth in order to anticipate a declining slope.

Fig. 6.1. Total electricity consumption in Great Britain, 1950–1985. x = year—1990; y = total consumption in terawatt hours (TWh).

It should have been obvious by 1965 that the reason for the steep rise in electricity demand was not the increase in total energy demand, which rose only 30% in the previous 15 years, but a replacement of other forms of energy by electricity in certain categories where it was a more convenient or cheaper source. As gas and oil would remain major competitors in heating and transport applications, it was unlikely that electricity demand would ever exceed a certain fraction of total energy demand. The curve fitted should therefore have an asymptote. At a more distant date other constraints would become obvious.

The simplest alternative curves are the logistic and Gompertz curves. The logistic curve fitted to the full data set gives a reasonable fit, although there are systematic patterns in the residuals, suggesting that errors are not independent, rather than that an alternative model should be found. A maximum demand of 252 TWh is predicted with a standard error of 6 TWh (assuming independent lognormally distributed errors). While this figure can be produced (with hindsight) after the asymptote can be seen to be approaching, what would have happened if the logistic curve had been used earlier in the series? Could this value have been forecast in 1965?

Unfortunately, we find that in 1965 the fitted logistic curve is indistinguishable from an exponential curve, with the estimated upper asymptote improbably large. It is not until 1970 that the decline in the increase is sufficient to make an acceptable forecast, an upper asymptote of 498 TWh with a standard error of 125 TWh. By 1972 the forecast is 377 ± 41 TWh, but thereafter the abrupt change in the data brings the forecast right down below 300 TWh.

One reason for the instability of the logistic curve is that the latest data point is very influential, especially while the slope is relatively large. While the fitted curve is close to the exponential fit, most of the information in the estimated asymptote is provided by that point. Furthermore, if a different curve such as the Gompertz curve is fitted, the curves are almost identical up to the latest data point, and diverge thereafter.

The choice of lognormal errors gives greater weight to the smaller observations, and as these occur at the beginning of the series it may be asked whether they are not given too much importance when they are furthest in time from the required predictions. There is always a temptation to include such points if they are consistent with the fitted curve, and to discard them if by doing so the fit is improved.

An alternative method of fitting is to regard the data as if they satisfy an underlying differential equation,

$$\frac{dy}{dt} = by\left(1 - \frac{y}{y_{\max}}\right),$$

but to estimate each yearly increase on the basis of the current values of y. Estimates of b and $y_{\max}$ are obtained by linear regression (through the origin) of $(y_{i+1} - y_1)$ and y_i and y_i^2. This method, known as *internal least squares* (Hartley, 1957) does not involve the time variable explicitly (unless the points are unequally spaced) but predicts $y_{\max}$ as the value of y for which dy vanishes.

The estimates of b and $\hat{y}_{max}$ are used to predict yearly increments, but their values may be revised as the new figures become available.

6.2.2. Car Ownership in the United Kingdom

Tanner (1978) and Brooks *et al.* (1978) discuss the problems of forecasting future road traffic densities so that adequate provision may be made. One component of the equation is the number of private cars per head of population, which has increased monotonically since 1951. Forecasts based on the logistic curve are discussed in both the cited papers. Tanner commented on a series of forecasts for 1975:

> Use of the logistic curve, coupled with the calibration methods employed, tended to forecast too rapid a growth in the early years.

In fact, the problem is very similar to that of the electricity series, that the data up to 1965 fit an exponential curve, and no evidence on a likely maximum value is available until the relative growth rate starts to decline. The data are given in Table 6.2 and illustrated in Fig. 6.2. Brooks *et al.* (1978) concentrated their discussion on the different predictions that result if the data are inverted. The logistic curve for "cars per head" implies an exponential curve for "persons per car." Using a nonlinear least squares procedure for both the direct and the inverted series they obtain the following estimates and 95% confidence limits for the asymptote of the logistic curve, for years 1951–1974.

	Lower	Estimate	Upper
Direct	0.26	0.33	0.40
Inverse	0.36	0.42	0.52

They comment that

> each estimate is outside the confidence interval based on the alternative model. It would therefore be somewhat naive to quote either confidence interval as if it represented the size of possible values consistent with the data. By changing the model only slightly, we immediately see just how misleading such a statement will be.

However, it should be noted that the difference is more than "slight," because the effect of inverting the data and assuming equal weighting is the same (approximately) as weighting in proportion to y^{-4}. Since the ratio of values in 1974 and 1951 is over 5, the inverse model gives the 1951 value 600 times the weight of the 1974 value. A more reasonable weighting is to assume lognormal errors, in which case it makes no difference whether the direct or the inverse model is fitted, and the estimates are 0.32, 0.36, and 0.43, respectively.

Table 6.2. Cars per head of population, United Kingdom

Year	Cars/head	Year	Cars/head	Year	Cars/head
1951	0.0487	1963	0.1416	1975	0.2534
1952	0.0511	1964	0.1574	1976	0.2586
1953	0.0561	1965	0.1692	1977	0.2587
1954	0.0628	1966	0.1796	1978	0.2589
1955	0.0712	1967	0.1935	1979	0.2681
1956	0.0781	1968	0.2020	1980	0.2764
1957	0.0837	1969	0.2088	1981	0.2793
1958	0.0905	1970	0.2137	1982	0.2868
1959	0.0982	1971	0.2231	1983	0.2918
1960	0.1084	1972	0.2344	1984	0.3008
1961	0.1164	1973	0.2482	1985	0.3082
1962	0.1265	1974	0.2506		

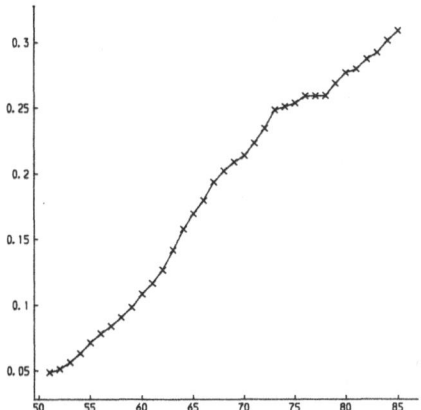

Fig. 6.2. Car ownership in Great Britain, 1950–1984. x = year—1900; y = number of cars per head of population.

6.3. Fitting Data to Models Defined by Differential Equations

Models defined by differential equations, such as

$$\frac{dy}{dt} = \beta y \left(1 - \frac{y}{\gamma} \right),$$
$$y(t_0) = \alpha,$$

(6.3.1)

occur in many practical contexts where rates of change are related to other variables and unknown parameters. There is no conceptual difference between such models and the explicit solutions such as equation (2.2.15)

$$E(y) = \frac{\gamma}{1 + \exp(-\beta(t - \mu))}$$

which is the logistic curve and is also the solution of (6.3.1). The main practical difference is computational: numerical integration of differential equations is inevitably much slower than evaluation of a function. The other difference is in the parametrization, and our ability to transform the parameter space, or to take advantage of separable linear parameters (Section 5.4). However, the choice of the initial time origin, t_0, may be made to coincide with a data value, which helps to ensure stability, and this may make the differential equation form easier to stabilize than the explicit form. Differential equation models involving external data variables cannot have explicit solutions and numerical methods must be used.

Since numerical solutions are lengthy to compute, and small changes in the parameters may produce large errors which build up rapidly, even to the point of computational overflow, it is important to have good initial estimates which may be obtained by using approximate methods which take less time and avoid overflow.

6.3.1. Approximate Method of Fitting Differential Equations Using Data Values for Expectations

A simple but effective technique for obtaining initial parameter estimates to models of the form

$$\frac{dy}{dt} = f(y, \theta), \qquad y(t_0) = \alpha,$$

is to assume that a sufficiently good fit exists so that little accuracy will be lost in replacing the expected values in $f(y, \theta)$ by the observed values in the data, using interpolation if necessary. Given trial values of θ, evaluate

$$E(y_i') = f(y_i, \theta),$$

$$E(y_i) = \int_{t_0}^{t_i} E(y_i') \, dt + \alpha,$$

using numerical integration based on the data values alone. The simplest method is the trapezoidal rule, but quadratic or cubic methods may be used even when the values of t are not evenly spaced. The log-likelihood or residual sum of squares may then be computed in the usual way and optimized with respect to θ.

Usually, there is at least a linear multiplier within the expression $f(y_i, \theta)$, if only to express the different dimensionality of y and t, and so for models of the form

$$\frac{dy}{dt} = \beta f(y, \theta),$$

the parameters β and α are separable (in the case of normal errors), and the solution may be obtained by optimizing with respect to nonlinear parameters only.

The full solution, in terms of all the parameters using accurate methods of integration such as the fourth-order Runge–Kutta method applied to sub-intervals of the differences between successive data times, may now be computed. The more subintervals are used the more accurate the results, but at the expense of computing time; in the context of statistical data fitting it is rarely worthwhile expending too much effort on excessive accuracy.

The approximate method is often adequate so far as the parameter estimates are concerned, but the dispersion matrix of parameter estimates is incorrect because important terms are neglected by fixing the trial values of $E(y)$ in the expression for $E(y_1^i)$.

6.3.2. A Practical Example in Plant Pathology

Buwalda *et al.* (1982) describe experiments in whcih the spread of fungi (vesicular arbuscular mycorrhizal infections, VAM) is measured in the root systems of growing plants. Models were studied in which the length of infected root at time, $t(y)$, was related to the total length of root, L, some of which was uninfected, and some of which might never become infected. One model which fitted reasonably well was the differential equation

$$\frac{dy}{dt} = \beta y \left(1 - \frac{y}{(\gamma L)} \right), \tag{6.3.2}$$

which, of course, is the same as (6.3.1) except for the variable L which is an external measurement whose form is not modeled explicity. The model is in effect a logistic with a moving asymptote, expressing the fact that the fungus grows both because of the supply of infective sources (y), but also because of

Table 6.3. Spread of fungal infection in roots of white clover (Buwalda *et al.*, 1982).

Days elapsed	Total root length (cm)	Infected root length (cm)
$t =$ 0	$L =$ 3.8	$y =$ 1.31
9	17.7	3.8
17	77.6	16
29	161	45.9
38	327	112
46	692	305
55	1504	744
65	2465	1442
75	4119	2475
85	7896	5131
95	11591	8005

the existence of a proportion of the uninfected root being within reach of infection.

An example of a data set is given in Table 6.3. The plant is white clover (*Trifolium repens*), and sample plants were dug up and examined at approximately ten-day intervals. Because the data were not evenly spaced the integration formulas used were more complex than those for equally spaced data. The parameter γ must be less than 1 and is approximately y/L for large t, and so it was not difficult to obtain the approximate solution by numerical integration of the expression

$$Z_i = \int_0^{t_i} y_i \left(1 - \frac{y_i}{(\gamma L_i)}\right),$$

and estimate $\alpha = E(y(t_0))$ and β by linear regression of y_i on Z_i. As is usual with growth data the weighting assumed lognormal errors, and the approximate solution obtained was

$$\gamma^* = 0.859,$$

$$\beta^* = 0.181,$$

$$\alpha^* = 1.395.$$

From this starting value the model was fitted again using the Runge–Kutta method, interpolating for L as necessary at intermediate points between data values, to obtain the estimates with their standard errors and correlations, as follows:

	Estimate	S.E.	Correlations		
$\hat{\gamma}$ =	0.879	± 0.074	1		
$\hat{\beta}$ =	0.168	± 0.010	−0.78	1	
$\hat{\alpha}$ =	1.312	± 0.158	0.17	−0.41	1

In the full study several species and treatments were compared, and parallel model analysis was used to test whether some parameters could be common to all data sets.

Because of the biological meaning attached to the parameters as specified and the reasonable results obtained there was no need to look for more stable parameters in this example. But in other cases, especially where convergence is difficult to achieve, perhaps because of insufficient data covering the entire course of growth, it would be advantageous to use empirical reparametrizations likely to reduce the correlations.

CHAPTER 7

A Program for Fitting Nonlinear
Models, MLP

The ideas described in this book have been developed and tested using the computer program, the maximum likelihood program (MLP). The aim of the MLP is to bring nonlinear modeling within the scope of the applied scientist by removing the computational problems of optimization to background status, and concentrating on the essential statistical analyses and the computation of relevant quantities. For example, the scientist wishing to find the best nonlinear curve to fit some data is usually more concerned with the shape of the curve (the fitted values) than with the estimates of parameters themselves.

7.1. Structure of MLP

The essential computational steps required in nonlinear model fitting have been described in Section 5.2, and the detailed description of the program language may be found in the current MLP manual (Ross, 1987).

The MLP has its own command language for data manipulation, selection, and fitting of models, and diagnostic output. The user prepares one or more data sets and then asks for one (or more) models to be fitted. A menu-driven version is being developed (Berzuini *et al.*, 1986) in which aspects of expert systems are to be introduced.

Models are classified into broad fields of application (curve fitting, assay analysis, discrete distributions, continuous distributions, or general user-defined models) within which particular models may be selected or defined. Within a common set of rules, the data structure is defined for each class of models which allows data to be transformed, generated, displayed, or otherwise manipulated in a standard way.

The main data structure consists of a set of v variates each of n units which may be subdivided into s subsets of size $n_1, n_2, \ldots, n_s$. The remaining data store is addressed in terms of v and n, giving space for fitted values, working variates, and other intermediate results. Other data areas are designated for scalar quantities, parameters, and compiled instructions which specify models and transformations. For each standard application a particular meaning may be assigned to each data structure, so that the default specification of a job is as simple as possible. The existence of subsets gives a convenient framework for *parallel model analysis*, the fitting of a model to each subset followed by a sequence of fits with some or all parameters constrained to be common to all subsets.

7.1.1. Computation of Log-Likelihoods

The relative log-likelihood is computed from the observation variate (y), the expectation variate $(z(\theta))$ or components $(z_1(\theta), z_2(\theta), \ldots)$, and the weight variate (w) which is by default assumed to take the value 1. Direct computation of the log-likelihood is only possible for ungrouped samples from frequency distributions (see Section 1.1.4), and is used as a refinement to a previous fit to grouped data. Where possible linear parameters are estimated separably. The standard error distributions and their respective log-likelihoods, $L(\theta)$, are as follows:

(1) Normal: $L(\theta) = \sum w(y - z)^2/2\sigma^2$, with variance parameter σ^2 estimated separately.

(1a) Normal, with separable linear parameters: $L(\theta) = $ weighted residual sum of squares from linear regression of y on $z_1(\theta), z_2(\theta)$, with or without a constant term.

(2) Lognormal: if $\log(y + c)$ is assumed normally distributed, for an appropriate choice of c, formula (1) is used. As a first stage, to take advantage of separable linear parameters, weights $(y + c)^{-2}$ are used with formula (1a). The exact solution using formula (1) may then follow.

(3) Poisson: y must be nonnegative and z must be strictly positive. Then $L(\theta) = \sum(y \log(y/z) + z - y)$.

(3a) Poisson, with separable scale parameter: when the expectation of y is $\beta z_1(\theta)$ then $\hat{\beta} = \sum y/\sum z_1$ is substituted for β in formula (3), which simplifies since $\sum z = \beta \sum z_1 = \sum y$.

(4) Binomial: assuming that $y = p$ is the proportion out of sample size $w = n$, so that wy is the observed count, and z is the expected proportion,

$$L(\theta) = \sum w\left(y \log\left(\frac{y}{z}\right) + \frac{(1 - y) \log(1 - y)}{1 - z}\right).$$

The formula may also be written in terms of $r = wy$.

(5) Multinomial: assuming that the elements of y are nonnegative counts, and that z is the corresponding expected probability such that $\sum z = 1$,

then $L(\theta) = \sum y \log(y/nz)$ where $n = \sum y$. If $\sum n \neq 1$ the formula is equivalent to (3a). More constraints may be imposed on z, resulting in fewer degrees of freedom but no change in the form of $L(\theta)$.

(6) Gamma: if both y and z are strictly positive, then the dispersion parameter of the gamma distribution is separately estimable, and the log-likelihood is proportional to

$$L(\theta) = \sum \left(\log\left(\frac{z}{y}\right) + \frac{y}{z} - 1 \right).$$

(6a) Gamma, with separable scale parameter: if $z = \beta z_1$, then $\beta = \sum (y/z_1)/n$ which is then substituted into formula (6).

(7) Negative binomial: y must be nonnegative, and a dispersion parameter k must be specified. Then

$$L(\theta) = \sum \left(y \log\left(\frac{y}{z}\right) - \frac{(y + k) \log(y + k)}{z + k} \right).$$

(8) Direct: if $z = f(\theta)$ is a probability density, $L(\theta) = -\sum \log f(\theta)$.

In all cases except (8) the minimum value, attained when $z = y$, is zero. $L(\theta)$ is the relative log-likelihood, and its minimized value multiplied by two (the residual deviance) is distributed asymptotically as χ^2, and is hence a value comparable to the error degrees of freedom. For models with a scale parameter ((1) and (6)) the residual deviance is of the order of magnitude of this scale parameter. The program therefore can assume that the choice of step lengths, convergence criteria, and output formats will be roughly appropriate for most data sets, if the model chosen is reasonable.

7.2. Curve Fitting $y = f(x, \theta) + \varepsilon$

The curves fitted by MLP are those commonly required in practical applications. In their general form they are independent of scale and origin, though one constraint is usually permitted, such as that the origin or an asymptote be fixed. Assuming normally distributed errors they may all be expressed in terms of separable linear parameters (Section 7.1.1, Method 1(a)), with an optional weight variate which may be constructed to handle lognormally distributed errors (Method (2)).

To specify the analysis the user must select:

(1) *Model formula* from

(a) polynomials up to the fifth degree;
(b) splines with one node;
(c) exponential curves:

$$y = \alpha + \beta \exp(-kx) = \alpha + \beta\rho^x,$$

$$y = \alpha + \beta_1 \exp(-k_1 x) + \beta_2 \exp(-k_2 x),$$
$$y = \alpha + (\beta + \gamma x) \exp(-kx),$$
$$y = \alpha + \beta \exp(-kx) + \gamma x;$$

(d) two-compartment curves:

$$y = \alpha \frac{\lambda \exp(-\mu x) - \mu \exp(-\lambda x)}{\lambda - \mu},$$

$$y = \alpha \lambda \frac{\exp(-\mu x) - \exp(-\lambda x)}{\lambda - \mu};$$

(e) ratios of polynomials:

$$y = \alpha + \frac{\beta}{1 + \delta x} \qquad \text{(rectangular hyperbola)},$$

$$y = \alpha + \frac{\beta}{1 + \delta x} + \gamma x,$$

$$y = \alpha + \frac{\beta + \gamma x}{1 + \delta x + \varepsilon x^2};$$

(f) generalized hyperbola:

$$y = \alpha + \frac{\beta}{(1 + \gamma x)^\delta};$$

(g) growth curves:

$$y = \alpha + \frac{\gamma}{1 + \exp(-\beta(x - \mu))} \qquad \text{(logistic curve)},$$

$$y = \alpha + \frac{\gamma}{(1 + \tau \exp(-\beta(x - \mu)))^{1/\tau}},$$

$$y = \alpha + \gamma \exp(-\exp(-\beta(x - \mu))) \qquad \text{(gompertz curve)};$$

(h) Gaussian curves:

$$y = \alpha + \frac{\beta}{\sqrt{2\pi\sigma^2}} \exp\left(-\frac{(x - \mu)^2}{2\sigma^2}\right),$$

$$y = \frac{1}{\sqrt{2\pi\sigma^2}} \left(\beta_1 \exp\left(-\frac{(x - \mu_1)^2}{2\sigma^2}\right) + \beta_2 \exp\left(-\frac{(x - \mu_2)^2}{2\sigma^2}\right)\right);$$

(i) Fourier curves:

$$y = \alpha + \beta \sin\left(\frac{2\pi(x - \varepsilon)}{w}\right),$$

$$y = \alpha + \beta_1 \sin\left(\frac{2\pi(x - \varepsilon_1)}{w}\right) + \beta_2 \sin\left(\frac{4\pi(x - \varepsilon_2)}{w}\right).$$

Details of the method of fitting are given below.

(2) *Constraints*
One degree of freedom may be sacrificed by constraining:

(a) the origin, fixing the value of y when $x = 0$ (1(a)–1(e));
(b) the asymptote, fixing the value of y when x is infinite (1(c), 1(e)–1(h));
(c) the period w of a Fourier curve (1(i));
(d) the node position of a spline curve (1(b)).

(3) *Error distribution*

(a) normal, unweighted;
(b) normal, with weight specified;
(c) lognormal;
(d) errors in both variables, with specified variance ratio.

Other error distributions such as the binomial are handled elsewhere in MLP.

(4) *Analysis of parallelism*
When the data matrix is partitioned into subsets each set is analyzed separately, followed by:

(a) fitting common nonlinear parameters;
(b) fitting common nonlinear parameters, then fitting a common curve with variable intercept (parallel displacements in the y direction);
(c) fitting a common curve with variable origin (parallel displacements in the x direction).

Finally, a common curve is fitted to all data together to complete the analysis of variance. When constraints are applied some intermediate stages may be omitted.

(5) *Error estimation:*

(a) from the residual mean square;
(b) from the within-replicates mean squares, if some x values are repeated;
(c) from a value supplied by the user (derived, e.g., from the analysis of variance in a designed experiment).

The standard output consists of a display of parameter estimates with their estimated standard errors and correlation matrix, followed by the residual sum of squares, degrees of freedom, and mean squares.
Additional optional output consists of:

(6) A table of observed and fitted values, weights, and weighted residuals. Relatively large residuals are marked with asterisks as indicating possible outliers.

(7) Standard errors of prediction, or their standardized squared reciprocals (known as effective replication) (Section 3.3). Small values of effective replication are asterisked as indicating influential points.

(8) Slopes of the curve at each data point, with standard errors if (7) is also requested.

(9) Derivatives with respect to each parameter, and parameter loadings (Section 3.3.2) (effects of each point on each parameter).

(10) Graphs of the data and fitted curve. If standard errors of prediction are available a confidence band is also plotted. If several subsets are analyzed the combined data sets, curves, and parallel curves are plotted. If effective replication is specified (7), it is also plotted as a separate graph. A residual plot is also available, with the difference between the current fitted curve and the previous fitted curve (see, e.g., Fig. 4.8).

Extra analyses are also available:

(11) *Predictions*
If further values of x are supplied, fitted values, slopes, etc., may be output in an extra table.

(12) *Inverse interpolation*
If further values of y are specified, values of x may be found (if they exist) such that $f(x) = y$. The estimated standard errors of the fitted $\hat{x}$ values assume that y is known exactly. If y is also subject to error the slope of the curve at $\hat{x}$ is given so that the approximate combined variance of estimated x may be obtained from the formula:

$$\text{variance of } \hat{x} = (\text{variance of } y \text{ fixed}) + \frac{\text{variance of } y}{(\text{slope})^2}.$$

In general, x is not necessarily a single-valued function of y, and if no solution is found within a moderate extension of the data range of x, the best value found is output, with no standard error.

(13) *Maxima and minima of the curve*
For models that include the possibility of a local maximum or a minimum, such as quadratic polynomials, mixed linear and exponential, or ratios of polynomials, the estimated values of x and y of the maxima and minima are given, together with their estimated standard errors.

(14) *Stable ordinates* (Section 2.2.3)
There exist sets of ordinates that are approximately uncorrelated (exactly uncorrelated for polynomial models), and the program attempts to find such a set by minimizing the sums of squares of the pairwise correlations between ordinates. The solution need not be unique (with polynomials a set can be found to include any given value of x), but the results provide a method for expressing the residual sum of squares in terms of approximately independent components.

(15) *Confidence limits on parameters*
Given the solution in stable ordinates (14), the Lagrangian algorithm (Section 5.5.3) may be applied to find the maximum and minimum value of each parameter, subject to the constraint $L(\theta) = $ constant, where the constant is the critical value of the log-likelihood for testing the hypothesis $\theta = \hat{\theta}$ for a given significance level. The program solves the simultaneous equations $f(\theta) = y$ by an iterative Newton method inverting the nonsymmetric matrix of the first derivatives with respect to the parameters. The solution will not exist if the confidence contour in parameter space is not closed.

7.2.1. Fitting the Curves

The program first checks the number of active data points (ignoring those with weight zero) which must be at least equal to the number of parameters fitted. If the error degrees of freedom are zero, the parameters are fitted but no standard errors are estimated (the fit should be exact if the data points conform to the model, otherwise a limiting case is fitted).

7.2.2. Fitting Polynomials

Polynomials are fitted noniteratively using the method of Forsyth (1957) adapted to allow for weighting and for the constraint that the curve should pass through a given point. The basic idea is that orthogonal polynomials $P_0(x)$, $P_1(k)$, ..., $P_k(x)$ of degree 0, 1, ..., k, respectively, are evaluated using the recurrence relationship

$$P_j(x) = (x - a)\, P_{j-1}(x) + b P_{j-2}(x), \qquad (7.2.1)$$

where

$$a = \frac{\sum w x P_{j-1}^2(x)}{\sum w P_{j-1}^2(x)}, \qquad (7.2.2)$$

$$b = \frac{\sum w P_{j-1}^2(x)}{\sum w P_{j-2}^2(x)}, \qquad (7.2.3)$$

and $P_0(x) = 1$ if the model is unconstrained, or $P_0(x) = 0$, $P_1(x) = x$, if the model is constrained.

Coefficients of x^j in each polynomial are built up using the values of a and b. The estimated parameters in the polynomials are obtained from the estimates of the orthogonal polynomial coefficients α_j in

$$E(y) = \sum \alpha_j P_j(x), \qquad (7.2.4)$$

$$\hat{a}_j = \frac{\sum w y P_j(x)}{\sum w P_j^2(x)}, \qquad (7.2.5)$$

and the variance of $\hat{a}_j$ is the residual mean square divided by $\sum w P_j^2(x)$. The analysis of variance gives the component sums of squares on one degree of freedom for each degree of polynomial fitted, which is $\hat{\sigma} \sum wy P_j(x)$. The dispersion matrix of the polynomial coefficients is not normally required, and it is sufficient to give the standard errors of the leading coefficients of each polynomial. Information required to compute standard errors of function is best obtained from the orthogonal polynomial coefficients.

Parallel curve analysis fits the models with displacement in x or y, but does not operate when the curves are constrained to pass through the origin.

7.2.3. Fitting Splines

Splines with one node are models of the form

$$E(y) = \beta_0 + \sum_{j=1}^{k} \beta_j x^j + H(x - \alpha) \cdot \gamma (x - \alpha)^k, \qquad (7.2.6)$$

where $H(t) = 0$ if $t < 0$ and $H(t) = 1$ if $t \geq 0$. In effect, the curve is one polynomial of degree k in the range $x < \alpha$ and a second polynomial of degree k in the range $x \geq \alpha$. The curve is differentiable $k - 1$ times at the node $x = \alpha$. There are $k + 3$ parameters to be estimated, of which α is nonlinear and the remaining parameters are linear given α. The model may be fitted only if there are at least $(k + 2)$ distinct data values of x on either side of α. The program therefore estimates a range of search for α based on the data range.

The residual sum of squares as a function of α is discontinuous in its derivatives of order k and over, because the function $H(x - \alpha)$ alters the model as α passes a data value of x. For α outside the data range the fit is that of a single polynomial of order k and the RSS is therefore constant. The RSS may have (a) more than one local minimum, (b) a local or global minimum at a data point, $\alpha = x_j$, and if $k = 1$ the first derivative is discontinuous (see, e.g., Fig. 4.22).

Estimates of standard errors of parameters are not valid because of the discontinuous nature of the model and the RSS function. The program therefore tests whether the fit is significantly better than the single polynomial of degree k, and if so, estimates numerically the confidence limits for $\hat{\alpha}$. Estimates of standard errors of prediction are similarly invalid and to establish confidence limits a separate analysis would be necessary for each data point.

Parallel curve analysis tests for equality in α between subsets, or for displacements in the x or y directions.

7.2.4. Fitting Exponential Curves

The exponential curves in MLP are solutions of simple differential equations

$$y' + k(y - \alpha) = 0,$$

$$y'' + (k_1 + k_2) y' + k_1 k_2 (y - \alpha) = 0, \qquad (7.2.7)$$

including some special cases of the latter. They have either one or two nonlinear parameters so that the main problem is to find a suitable transformation of those nonlinear parameters.

The single exponential,

$$y = \alpha + \beta \exp(-kx) = \alpha + \beta\rho^x, \qquad (7.2.8)$$

has already been discussed (see Fig. 2.4). Optimization of ρ rather than k has considerable advantage if x is suitably scaled, although there are problems when the fit is almost exactly a straight line. The user is asked to specify whether a positive or a negative exponential is expected; usually there is an underlying model which indicates one or the other, and if the fitted solution is of the wrong kind the results may not be interpretable.

To find an appropriate scaling for x, MLP converts x to have range $(0, 2)$, reversing the sign if a positive exponential is specified, so that the parameter ρ should lie in the range $(0, 1)$. If $\rho = 1$, the fit is to a straight line and if $\rho = 0$ the fit is to a step function, with the lowest value of x differing from all others (to which the overall mean is fitted). Between these extremes the shape of the scaled response depends on the extent to which observations are sampled on the nearly flat asymptotic portion of the curve, which is discovered by the following procedure: evaluate the RSS at $\rho = 0.1, 0.3, 0.5,$ $0.7,$ and 0.9 and select the lowest value; optimize with respect to ρ within the range $(0.02, 0.98)$; if the optimum is on either bound, rescale x so that the new value of ρ is 0.5, and try again; if the optimum is again on a bound, report an error message and fit the limiting case. This procedure may occasionally fail to identify correctly cases where valid solutions exist very close to the limits, either because of the design (distribution of x) or where the fit is extremely poor, but it works well in nearly every practical case.

Similar procedures are used for the models

$$y = \alpha + (\beta + \gamma x)\,\rho^x, \qquad (7.2.9)$$

$$y = \alpha + \beta\rho^x + \gamma x, \qquad (7.2.10)$$

where in the first case the working regressor variables are ρ^x and $x\rho^x$. This model (7.2.9) is the limiting case of the double exponential when the two exponential components tend to equality. Because there is a frequent possibility that the RSS with respect to ρ has a local minimum, the initial search along equally spaced values is important, but not totally reliable as a method for ensuring that the best fit is found. The second model (7.2.10) is easier to fit and usually has only one solution. In both cases the solution may be "out of bounds," and the limiting case as ρ tends to 1 is a quadratic polynomial.

The full double exponential curve

$$y = \alpha + \beta_1\rho_1^x + \beta_2\rho_2^x \qquad (7.2.11)$$

is optimized with respect to ρ_1 and ρ_2 within the unit square, using separate scale factors for each term. Initially, the scale factor is the same as for the single exponential curve, and a design of 15 trial values samples the RSS to find a suitable starting point. The scale factors are then adjusted so that the search can restart from $(0.5, 0.5)$ and after a few iterations the scale factors are adjusted again. This procedure helps to ensure that when a valid solution exists, the residual sum of squares function is a suitable shape for optimization. The original scale is then restored for output, and the following checks made: if ρ_1 and ρ_2 are very close in value, limiting case (7.2.9) should be fitted; if either ρ_1 or ρ_2 are very close to 1 (before rescaling) limiting case (7.2.10) should be fitted; if either ρ_1 and ρ_2 are very close to zero the limiting case is to ignore the lowest value of x and to fit a single exponential to the remaining data.

All the foregoing models may be fitted with constraints, as follows: if $y = \alpha$ when $x = 0$ (fixed origin) use the regression of $y - \alpha$ on $(\rho^x - 1)$, etc., without an intercept term; if $y = \alpha$ when x is infinite (fixed asymptote) use regression of $y - \alpha$ on ρ^x, etc. The optimization procedures are otherwise unchanged.

Parallel curve analysis is available in all three forms (Section 7.2(4)).

7.2.5. Fitting Two-Compartment Curves

The special cases of (7.2.11) arising from the two-compartment model

$$u \underset{\lambda}{\rightarrow} v \underset{\mu}{\rightarrow},$$

where $u' = -\lambda u$, $v' = \lambda u - \mu v$, and at time $x = 0$, $u = \alpha$, and $v = 0$, are:

(i) when $u + v = y$,

$$y = \frac{\alpha}{\lambda - \mu}(\lambda \exp(-\mu x) - \mu \exp(-\lambda x)); \qquad (7.2.12)$$

(ii) when $v = y$,

$$y = \frac{\alpha \lambda}{\lambda - \mu}(\exp(-\mu x) - \exp(-\lambda x)). \qquad (7.2.13)$$

The occurrences of λ and μ in more than one place in the formulas makes it difficult to transform the parameters, as has been discussed in Section 3.1.1. The procedure used assumes that data values of x are positive, and that both $\bar{x}$ and $2\bar{x}$ are close to some data values. The stable parameters which transform to λ and μ are approximately the ratios of ordinates at $x = 0$, $x = \bar{x}$, and $x = 2\bar{x}$ for model (7.2.12) and $x = \frac{1}{2}\bar{x}$, $x = \bar{x}$, and $x = 2\bar{x}$ for model (7.2.13). The scale parameter α is estimated by linear regression without an intercept, given λ and μ. The formulas used are as follows:

Model (7.2.12), approximately,

$$\phi_1 = \frac{y(\bar{x})}{y(0)}, \qquad \phi_2 = \frac{y(2\bar{x})}{y(0)}.$$

Then

$$\mu = \log\left(\frac{\phi_1}{\phi_2}\right), \qquad \lambda = \frac{\phi_1^2 \mu}{\phi_1^2 - \phi_2},$$

if $\phi_1^2 = \phi_2$, use the form

$$y = \alpha \exp(-\mu x),$$

if $\phi_1^2 < \phi_2$, use the form

$$y = \frac{\alpha}{\lambda - \mu}(\lambda \exp(-\mu x) - \mu \exp(\lambda x)).$$

Model (7.2.13), approximately,

$$\phi_1 = \frac{y(\bar{x})}{y(\frac{1}{2}\bar{x})}, \qquad \phi_2 = \frac{y(2\bar{x})}{y(\bar{x})}.$$

Then evaluate $c = 2\phi_2 - \phi_1^2$, and if $c > 0$,

$$\lambda = -\log \tfrac{1}{2}(\phi_1 - \sqrt{c}), \qquad \mu = -\log \tfrac{1}{2}(\phi_1 + \sqrt{c}),$$

but if $\phi_1 < \sqrt{c}$, use the form

$$\lambda = -\log \tfrac{1}{2}(\sqrt{c} - \phi_1),$$

$$y = \frac{\alpha}{\lambda - \mu}\left(\exp(-\mu x) - \exp(-\lambda x) \cos\left(\frac{\pi x}{\bar{x}}\right)\right),$$

and if $c = 0$,

$$y = \alpha x \exp(-\mu x) \qquad \text{where} \quad \mu = -\log(\phi_1/2),$$

and if $c < 0$,

$$\mu = -\frac{\tfrac{1}{2}\log(\phi_1^2 - \phi_2)}{1}, \qquad \gamma = \tan^{-1}\left(\frac{\sqrt{-c}}{\phi_1}\right),$$

and

$$y = \alpha \exp(-\mu x) \sin(\gamma x).$$

This complicated set of transformations in fact generates a continuous objective function across the critical boundaries where the square roots or logarithms would otherwise have negative arguments. When the solution is found errors are reported if the model is not of the required form.

7.2.6. Fitting Ratios of Polynomials

The rectangular hyperbola is written in the form

$$y = \alpha + \frac{\beta}{1 + \delta x} \qquad (7.2.14)$$

to resemble the exponential curve (7.2.8). Although δ is the nonlinear parameter printed by the program, it actually minimizes with respect to δ' where $y = \alpha + \beta'/(1 + \delta'(x - \bar{x}))$ and the defining parameters are recovered after convergence. So that the denominator must not change sign within the data range, bounds are placed on δ' related to the maximum and minimum values of x, namely,

$$\frac{0.99}{\bar{x} - x_{max}} < \delta' < \frac{0.99}{\bar{x} - x_{min}}.$$

These limits may be exceeded for a valid fit if the vertical asymptote is very close to x_{min} or to x_{max}, but more often it is because the "best" fit occurs where the vertical asymptote lies within the interval (x_{min}, x_{max}). A warning is given when this occurs.

The general nonrectangular hyperbola,

$$y = \alpha + \frac{\beta}{1 + \delta x} + \gamma x, \tag{7.2.15}$$

resembles the line + exponential (7.2.10) and is fitted in the same way in terms of the nonlinear parameter δ', with the same bounds on the range. The extra parameter brings the added danger of false local optima, especially where the plotted data are nearly linear; the false solutions correspond to hyperbolas with a common nonvertical asymptote (the linear fit) with vertical asymptotes close to one or more data points. These solutions are excluded in the first round of optimization, but may be found if no valid solution (without the discontinuity) exists.

The ratio of quadratics,

$$y = \alpha + \frac{\beta + \gamma x}{1 + \delta x + \varepsilon x^2}, \tag{7.2.16}$$

is a five-parameter model with two nonlinear parameters, and it has various forms, depending on whether the denominator has real or complex roots. If the roots are complex the curve is continuous, but if the roots are real there are two vertical asymptotes, and for practical usefulness it is desirable that they are both either less than x_{min} or greater than x_{max}. The flexibility of the model is such that numerous local optima are possible. MLP excludes false solutions in the first round of optimization by using parameters δ' and ε' in the equation

$$y = \alpha + \frac{(\beta' + \gamma' x)}{1 + \delta'(x - \bar{x}) + \varepsilon'((x - \bar{x})^2 - c)},$$

where $c = \sum (x - \bar{x})^2/n$.

Then if δ' has the same bounds as for the hyperbola, and

$$\frac{0.02}{c} < \varepsilon' < \frac{0.98}{c},$$

then it can be shown that the denominator is always positive between x_{min} and x_{max} so that the curve is continuous. Again, there are data sets for which the solution lies outside the bounds; whether or not these generate a warning message depends on the positions of the vertical asymptotes. The constraint that the asymptote is fixed ($\alpha = $ constant) generates the inverse linear (7.2.14) or the linear/quadratic (7.2.16), but has no particular use for (7.2.15). In all cases the curve can be constrained to pass through the origin or some other point, which gives curves equivalent to

$$y = \frac{\beta x}{1 + \delta x}, \frac{\beta x + \gamma x^2}{1 + \delta x}, \text{ and } \frac{\beta x + \gamma x^2}{1 + \delta x + \varepsilon x^2},$$

which are often written in other forms.

Parallel curve analysis in all three forms (common nonlinear parameters, displaced intercept, displaced origin) often shows that although the parameters fitted to individual data sets are widely different, because of the correlations between them, they may in practice represent very similar curves when plotted against the data.

7.2.7. Fitting the Generalized Hyperbola

The generalized hyperbola (Turner, 1962) of the form

$$y = \alpha + \frac{\beta}{(1 + \gamma x)^\delta} \tag{7.2.17}$$

was expected by its authors to introduce a wide range of curves which would give better fits than the exponential or rectangular hyperbola which are special cases, $\delta = 0$ and $\delta = 1$, respectively). Unfortunately, the shapes generated by variations in γ and δ are not sufficiently different to be of very great use except in large and accurate data sets. Note that $(1 + \gamma x)$ must be strictly positive for all data.

To find a stable parametrization for γ and δ the ideal solution is to use the standardized curve through the four points $(0, 0)$, $(1/3, \phi_1)$, $(2/3, \phi_2)$, $(11, 1)$, and to solve for γ and δ in terms of ϕ_1 and ϕ_2. As that is not computationally simple, an alternative procedure is used which generates estimates of γ and δ rather than their true values (in terms of the reverse transformation). The details are as follows:

(1) Fit an exponential curve in terms of parameter $\hat{\phi}_1$ alone.
(2) Given $\hat{\phi}_1$, evaluate $\hat{\phi}_2$ for the exponential curve and the rectangular hyperbola giving two reference values, q_e and q_h, respectively. Rescale $\hat{\phi}_2$ with respect to these values, estimating δ on a reciprocal scale and $\gamma\delta$ on a direct scale. Use the exponential function if $\hat{\phi}_2 = q_e$ exactly, and the logarithmic function if $\delta = 0$ exactly. Start the optimization from the exponential fit for $\hat{\phi}_1$ with corresponding $\hat{\phi}_2 = q_e - 0.002$.

This transformation enables the residual sum of squares to vary continuously across the boundary between positive and negative values of δ subject to $\gamma\delta$

being constant. The first stage is essential because there are local optima corresponding to unsuitable combinations of ϕ_1 and ϕ_2.

The model may be constrained to fix the origin or the asymptote in the same way as for the exponential and rational curves.

7.2.8. Fitting Growth Curves

A detailed discussion of the method of fitting the logistic curve,

$$y = \alpha + \frac{\gamma}{1 + \exp(-\beta(x - \mu))}, \qquad (7.2.18)$$

is given in Section 5.1.1. The method is to express β and μ in terms of ratios of ordinates at three equally spaced values of x, suitably located with respect to the data. The transformation is more reliable if α is known approximately, because if α is very poorly indicated by the data the ordinates on which stability is based will be themselves unstable.

The generalized logistic curve,

$$y = \alpha + \frac{\gamma}{(1 + \tau \exp(-\beta(x - \mu)))^{1/\tau}}, \qquad (7.2.19)$$

has three nonlinear parameters, and there is no simple way of solving for τ in terms of stable ordinates. The method used therefore is to treat τ separately and to use the fact that on a scale of $(y - \alpha)^{\tau}$ the curve becomes the ordinary logistic curve. The transformation chosen for τ is $u = \tau/(1 + \tau)$ which varies between 0 and 1 as τ varies from 0 to ∞. This is not a perfect choice for all data, as the difference in shape between the Gompertz curve ($\tau = 0$) and the logistic curve ($\tau = 1$) is less great than that between the logistic curve and the more extreme realization as τ tends to infinity, and it is possible that transformations such as $u = \tau/(\delta + \tau)$ (with δ different from 1) might be more effective in some cases. However, it works reasonably well. So the procedure is to start with $u = 0.5$ and to use the transformation for the logistic curve applied to the ordinates $(y - \alpha)^{\tau}$.

Several bounds may be encountered. If τ (or u) tends to 0 the Gompertz (7.2.20) is the limiting case. If u tends to 1 the limiting case is the "arrested exponential"

$$y = \alpha + \beta \exp(kx) \quad \text{if } y < y_{\max}, \quad \text{otherwise } y = y_{\max}.$$

Then, for any given τ the bounds for the ordinary logistic apply, the increase is slower than exponential, and the transition from one asymptote to the other is smooth.

The Gompertz curve,

$$y = \alpha + \gamma \exp(-\exp(-\beta(x - \mu))), \qquad (7.2.20)$$

may be defined as an exponential curve with respect to $\log(y - \alpha)$. The transformation of β and μ is made to depend on an initial estimate of α, and three ordinates at equally spaced values of x. Then, from the functions

$$\phi_1 = \log \log \left(\frac{y_2 - \alpha}{y_1 - \alpha} \right),$$

$$\phi_2 = \log \log \left(\frac{y_3 - \alpha}{y_2 - \alpha} \right),$$

it is possible to estimate β and μ, given the values of x. As with the logistic curve an initial choice of x depends on the distribution of the data x values. If $\phi_1 = \phi_2$ an exponential curve is fitted.

To any set of data a second form of the curve may be fitted, in which α is the maximum rather than the minimum value, and β is of opposite sign. Because of the basic asymmetric sigmoid shape both forms tend to fit the data about equally well, and a choice must be made as to which is required. The algorithms for the logistic and Gompertz curves require protection against overflow in the exponential terms, which may occur for outlying values of x. But since the basic nonlinear function is close to 0 or 1 in such cases, it is only necessary to test the relevant arguments and to substitute the appropriate value.

The constraint that the curve should pass through a particular point is not usually required, but the lower asymptote α may be fixed. The program also gives the upper asympote $(\alpha + \gamma)$ with its standard error. The value $x = -\infty$ sometimes needs to be included in the data, as in biological assays when x is on a log scale and $\log(0)$ is required. This special value may be indicated without involving the parameters β and μ.

Parallel curve analysis is particularly useful in testing whether a common shape of growth curve applies to individuals which attain a different ultimate size.

7.2.9. Fitting Gaussian Curves

An important application of fitting Gaussian curves,

$$y = \alpha + \frac{\beta}{\sqrt{2\pi\sigma^2}} \exp\left(-\frac{(x - \mu)^2}{2\sigma^2} \right), \qquad (7.2.21)$$

is to estimate β, the total area under the curve above the asymptote α, which is frequently constrained to zero. In various forms of spectrography β is an estimate of the quantity of a substance in a mixture; provided the various components do not overlap too closely, at most two at a time need be fitted.

The single Gaussian is fitted in terms of the nonlinear parameters μ and σ using as initial estimates the weighted mean and standard deviation of the sample x, using the sample y as a weight. Although this is not a correct procedure if the values of x are not equally spaced or the curve is incomplete, estimation usually presents no problem. The only possible source of trouble is when less than half of the curve is supplied and there is no maximum within the data range, in which case a false solution with negative β exists corresponding to an upside down Gaussian.

The corresponding double Gaussian form of the problem, with common σ^2 but different μ and β, needs more care. The single Gaussian is fitted first, and from its fitted parameters, $\hat{\mu}_0$ and σ_0, the initial estimates for the three nonlinear parameters μ_1, μ_2, and σ are assigned as follows:

$$\mu_1^{(0)} = \hat{\mu}_0 - 0.4\sigma_0,$$

$$\mu^{(0)} = \hat{\mu}_0 + 0.4\sigma_0,$$

$$\sigma^{(0)} = 0.6\sigma_0 \quad \text{with upper limit } 0.99\sigma_0. \tag{7.2.22}$$

False local optima may be found, in which one of the β parameters is negative, but this is unlikely in the context of spectral data when the above initial value procedure is used.

7.2.10. Fitting Fourier Curves

For data which are periodic, either exactly, because they relate to a natural cycle such as the day or the year, or approximately, because fitted values of y repeat themselves at intervals of w in x, where w is a parameter to be estimated, models of the form

$$y = \alpha + \beta \sin\left(\frac{2\pi(x - \varepsilon)}{w}\right),$$

$$y = \alpha + \sum_k \beta_k \sin\left(\frac{2k\pi(x - \varepsilon_k)}{w}\right), \tag{7.2.23}$$

are Fourier curves of order k. MLP fits either $k = 1$ or $k = 2$, with four or six parameters, respectively, which is adequate for most data.

There is only one nonlinear parameter, w, since the parameters β_k and ε_k may be obtained from the linear regression coefficients of $\sin(2k\pi x/w)$ and $\cos(2k\pi x/w)$. If w is to be estimated it should be noted that if w is small compared with the range of x, there will be several cycles with very few data points per cycle and spurious solutions may be found. MLP therefore starts by default with w equal to the range of x to find a solution with as few cycles as possible.

More often, the value of w is fixed at the natural period appropriate to the data.

7.2.11. Curve Fitting: The Subsequent Analyses

After fitting the curve by weighted least squares by the methods outlined in Sections 7.2.2–7.2.10, some of the available options specify further analyses requiring optimization. These are as follows:

(1) Lognormal errors

Having obtained the approximate parameter estimates using weights proportional to $1/(y + c)^2$, where c is a value assigned by the user, MLP may proceed to direct optimization of

$$\sum (\log(y + c) - \log(\hat{y} + c))^2,$$

which gives the correct estimates based on lognormally distributed errors. The difference in the fits is usually very small unless the solution depends heavily on small values of y and the fit is poor. On rare occasions a critical boundary may be crossed such that a valid solution exists with weighted least squares but not with log least squares.

(2) Errors in both variables

The functional relationship model (see, e.g., Sprent (1969), for the linear case) may be fitted if the ratio of the variances of x and y is specified in advance. Then a new variate $\hat{x}$ is created which is re-estimated for each trial value of the parameters, θ. If the variance of x is $(1 - \lambda) \sigma^2$ and the variance of y is $\lambda\sigma^2$, then $\hat{x}$ is estimated by iterative application of the formula

$$\hat{x}^{(1)} = \hat{x}^{(0)} + \frac{\lambda(x + \hat{x}^{(0)}) + (1 - \lambda) \hat{y}'(y - \hat{y})}{\lambda + (1 - \lambda)(\hat{y}')^2},$$

where

$$\hat{y} = f(\hat{x}^{(0)}, \theta) \qquad \text{and} \qquad \hat{y}' = \frac{df}{dx}(\hat{x}^{(0)}, \theta).$$

Then the objective function optimized is

$$\sum (\lambda(x - \hat{x})^2 + (1 - \lambda)(y - \hat{y})^2).$$

The minimized value, if $\lambda(1 - \lambda) \neq 0$, is divided by $\lambda(1 - \lambda)$ to give the "equivalent sum of squares," which tends to the same value as the ordinary "errors in y" model as λ tends to 1.

    The success of the method depends on the appropriateness of the functional relationship model to the particular data. The curve should be locally monotone with appreciable slope: the largest adjustments in x occur where the slope is least. Clearly, if the curve has maxima or minima then x will have more than one expectation $\hat{x}$ to choose from: the computing procedure simply assigns the nearest value on the fitted curve.

(3) Stable ordinates and support limits

The direct approach to finding a set of p uncorrelated ordinates, given the fitted model $f(\theta)$ and the dispersion matrix of parameters $V\sigma^2$, is to define an optimization problem for $\xi_1, \xi_2, \ldots, \xi_p$ which are p values of x. The objective function to be minimized is most simply the sum of squares of off-diagonal elements of $J'VJ$, where J is the Jacobian matrix $\partial f(\xi_i)/\partial \theta_j$. The choice of the covariance matrix rather than the correlation matrix makes little difference, because the diagonal elements of $J'VJ$ usually do not vary greatly in magnitude, and the solution of either achieves the same objective. It does not particularly matter if a minimum is not found, provided the correlations are small.

Given the solution in terms of ξ_i then the ordinates $\phi_i = f(\xi_i, \theta)$ may be treated as approximately orthogonal parameters, and the residual sum of squares function is approximated by

$$L(\phi) = \sum (\phi_i - \hat{\phi}_i)^2 (J'VJ)^{-1},$$

and the algorithm (described in Section 5.5.3) is used to find maxima and minima of each θ_i subject to the constraint $L(\phi) = c$, for the appropriate critical value of c. To solve the nonlinear simultaneous equations transforming from ϕ to θ iterative use is made of the adjustment

$$\theta^{(1)} = \theta^{(0)} + J^{-1}\phi(\theta^{(0)}).$$

The iteration will diverge if the confidence contour crosses a critical boundary, such that parameters become infinite or nonestimable.

(4) Parallel curve analysis

Individual estimates of θ for each subset are stored in a table along with the sums of squares and degrees of freedom. To fit common nonlinear parameters the transformations (described in Section 7.2.1) are no longer available because the data designs may differ (i.e., different x values). But the extra information from the subsets makes it relatively easier to optimize the total within-subset residual mean squares as a function of the nonlinear defining parameters. The initial values used are the means of the corresponding parameter estimates for the individual sets. If difficulties with convergence are encountered it is usually because one or more of the individual sets are very much out of line with the remainder.

To fit a common model apart from "vertical" displacement, or individual values of the "constant" or intercept term, the objective function in terms of the nonlinear parameters is, as for parallel linear regression, obtained by correcting each sum of squares or products for the subset means, and totalling the resulting components before obtaining the residual sum of squares by fitting the other linear parameters. Thus, if the model is

$$y = \beta_0 + \sum \beta_j z_j(\theta),$$

the jth sum of squares, for example, is $S_{jj} = \sum_k \sum_i (z_{jki}(\theta) - \bar{z}_{jk})^2$ and on convergence $\hat{\beta}_{0k} = \bar{y} - \sum \hat{\beta}_j \bar{z}_{yk}$.

To fit a common model with "horizontal" displacement (variable origins of x) a temporary variate $(x - \alpha_j)$ is created for each subset except the first, and the parameters α_j and θ are estimated alternately until the α_i converge. Initial estimates of α_j are obtained by parallel line analysis (on the assumption that the model is only appropriate for monotone data with curves of appreciable slope). Given the fitted values of θ, adjustments to α_j are then computed by a Gauss–Newton step

$$\alpha_j^{(1)} = \alpha_j^{(0)} + \frac{\sum w \frac{\partial \hat{y}}{\partial x}(y - \hat{y})}{\sum w \left(\frac{\partial \hat{y}}{\partial x}\right)^2}.$$

Speed of convergence depends on the slope of the curve at the data points, and clearly if the curves are very flat there is little information on α_j.

To complete the analysis of variance a common curve is fitted to all data. If the data sets are widely separated it may be difficult to obtain a fit, although it is usually obvious in such cases that a common curve is not appropriate, and that positional differences are significant.

7.3. Fitting Frequency Distributions

MLP fits discrete and continuous univariate frequency distributions to samples that are either unclassified or classified into groups. The distributions chosen meet a range of practical needs, but are not intended to be comprehensive. The classification of data and models is as follows:

(1) Discrete distributions (counts of 0, 1, 2 events, etc.) or continuous distributions.
(2) Classified data (number of observations in each of several classes with specified boundaries) or unclassified (single observations).
(3) Complete or truncated distributions (unknown numbers of observations in some classes).

The program operates on classified data even when the actual values are known. Appropriate class intervals must be supplied such that there is little loss of information on grouping. The necessary calculations are then much fewer, the parameters are easier to estimate, and a goodness-of-fit test may be applied. If required, the exact maximum likelihood estimates (MLEs) may follow in a secondary analysis, using the approximate estimates as initial values. The program first checks that the class limits are correctly ordered, and in the continuous case that no class contains too high a proportion of the observations. Sample moments and percentiles are then evaluated, which are often required for stable parametrizations.

When several samples are available simultaneously, a test of homogeneity of parameters is available when the same class intervals are chosen in each case. If there are k samples with c classes and p parameters are fitted to each set, and also to the totals over all samples an analysis of chi-squared is displayed as follows:

Source	Degrees of freedom	Comment
Samples heterogeneity	$(k-1)(c-1)$	Independent of fitted distribution
Between fits	$p(k-1)$	
Within samples	$k(c-p-1)$	Pooled from individual fits

The middle line is obtained from the residual chi-squared for the totals (on $c-p-1$ degrees of freedom) plus the top line minus the bottom line.

7.3.1. Fitting Discrete Distributions

Discrete distributions are generated by several types of process, as described by Johnson and Kotz (1972). Counts of the number of random events in a unit interval follow the Poisson distribution or one of its generalizations, the negative binomial, the Neyman Type A, the Polya–Aeppli, the Poisson lognormal, and the Poisson–Pascal. Counts of waiting times, or numbers of discrete intervals between events, follow the geometric distribution or one of its generalizations, which again includes the negative binomial under a different derivation. A third group of discrete distributions, with zero class missing, refer to counts of the number of occurrences in a sample of items of different classes, such as insects of different species caught in a trap; the logarithmic series distribution is the basic one-parameter model for such data.

Stable parameters depend on the data; for moderately skewed data mean and variance are fairly stable, but if the zero cell predominates then the expected proportion of zeros is more stable than the variance. If the zero cell is missing it is the moments of the truncated distribution that are stable rather than the moments of the complete distribution.

The methods actually used in MLP are as follows:

(1) *Poisson distribution*:

$$p_0 = \exp(-m), \qquad p_r = mp_{r-1}/r, \qquad r \geq 1.$$

Optimization is with respect to m starting at $m = \bar{r}$. Even when zeros are missing convergence is rapid.

(2) *Geometric distribution*:

$$p_0 = p, \qquad p_r = (1-p)p_{r-1}, \qquad r \geq 1.$$

Optimization is with respect to the expected mean, m starting with $\bar{r}$ gives $\hat{p} = 1/(1 + \hat{m})$ when zeros are included, and $\hat{p} = 1/\hat{m}$ when they are missing.

(3) *Logarithmic series distribution*:

$$p_r = (\theta^r/r) \log\left(\frac{1}{1-\theta}\right), \qquad r \leq 1.$$

Since $\hat{\theta}$ is usually close to 1, optimization is in terms of $z = \log(1/(1 + \theta))$ where the initial estimates of z is the solution of the implicit equation

$$z = \log(1 + z\bar{r}),$$

which may be solved by a simple iterative algorithm.

(4) *Negative binomial distribution*:

$$p_0 = (1 + m/k)^{-k}, \qquad p_r = \frac{m}{m+k} \frac{k+r-1}{r} p_{r-1}, \qquad r \geq 1.$$

For complete data the theoretical mean and variance, (m, v), are used as parameters, and the transformation $k = m^2/(v - m)$ is used except when $v = m$ when the Poisson distribution is fitted. When k is negative the formulas for p_r are positive up to $r = |k|$, so for data for which v is slightly less than m a positive binomial distribution is fitted.

When the zero cell is missing the program distinguishes between "Poisson-like" samples (with large k) and "log-series-like" samples with small k using the sample moment discriminant, $v_0 \gtrless m_0(m_0 - 1)$. When $v_0 < m_0(m_0 - 1)$, an estimate of the zero frequency is obtained from m_0, v_0, and the sample estimate of p_1

$$p_0^* = \frac{vp_1}{m^2(1 - p_1)}.$$

Parameters m and k are then obtained from m^*, v^*, and p_0^* using the formula

$$m = m^*(1 - p_0^*),$$

$$v = v^*(1 - p_0^*) + m^{*2}p_0^*(1 - p_0^*),$$

$$k = \frac{m^2}{v - m}.$$

When $v_0 \geq m_0(m_0 - 1)$ the parameters fitted are k and $z = \log(1 + m/k)$ where z is analogous to the log series parameter in section (3) above.

(5) *Neyman Type A distribution*:

$$p_0 = \exp(-m_1(1 - e^{-m_2})), \qquad p_r = \frac{m_1 m_2}{r} e^{-m_2} \sum_{j=0}^{r-1} \frac{m_2^j}{j!} p_{r-j-1}, \qquad r \geq 1.$$

This distribution was discussed in Section 2.1.1 where the effectiveness of the mean–variance transformation was illustrated. The basic formulas are

$$m_2 = \frac{v - m}{m}, \qquad m_1 = \frac{m}{m_2},$$

from which the probabilites are computed as functions of m and v. When the zero cell is missing the limiting distribution has variance $v^* = m^*(1 + 0.2985m^*)$, so that if v^* is too large no fit is attempted. Otherwise, the parameters are obtained using an estimate of p_0^* obtained from p_1 by the formula

$$p_0^* = p_1 \exp\left(\frac{v_0^*}{m_0^*} + m_0^* p_1 - 1\right) m_0^*,$$

and proceeding as for the negative binomial distribution.

(6) *Polya–Aeppli distribution*:

$$p_0 = e^{-m_1}, \qquad p_r = \frac{m_1 p}{r} \sum_{j=0}^{r-1} (r - j)(1 - p)^{r-j-1} p_j, \qquad r \geq 1.$$

This distribution lies between the negative binomial and the Neyman Type A, and similar methods are used to fit it. The mean variance transformation gives

$$p = 2m(m + v), \qquad m_1 = mp,$$

and the probabilities computed by a single recurrence formula. If $v < m, p > 1$ but positive values of p_r are computed for small r.

When the zero cell is missing the appropriate test is that v^* must not exceed $m^*(m^* - 1)$, and p_0 is estimated from

$$\frac{p_0^* p_1^* (v_0^*/m_0^* + p_1^* m_0^* + 1)^2}{4m_0^*}.$$

(7) *Poisson lognormal*:

$$p_r = \frac{1}{r!\sqrt{2\pi\sigma^2}} \int_0^\infty e^{-z} z^{r-1} \exp\left(-\frac{(\log z - \mu)^2}{2\sigma^2}\right) dz.$$

This distribution is derived from the assumption that observations are sampled randomly from Poisson distributions whose means are distributed lognormally with parameters μ and σ^2. The defining parameters μ and σ^2 are obtained from working parameters m and v by the formulas

$$\sigma^2 = \log\left(1 + \frac{v - m}{m^2}\right),$$

$$\mu = \log(m) - \tfrac{1}{2}\sigma^2.$$

The probabilities p_r are then evaluated by numerical integration.

When the zero cell is missing there is no simple way of estimating p_0^*, but in practice it is sufficient to use p_1. The quantity p_0^* is only a transformation constant designed to reduce correlation between m and v.

(8) *Poisson–Pascal*:

$$p_0 = \exp(-\lambda(1 - (1 + \pi)^{-k})),$$

$$p_r = \lambda \pi k (1 + \pi)^{-k-1} \left(p_{r-1} + \frac{\pi}{1 + \pi} (k + 1) p_{r-2} \right.$$

$$\left. + \sum_{j=2}^{\infty} \left(\frac{\pi}{1 + \pi} \right)^j \binom{k + j}{j} p_{r-j-1} \right).$$

This three-parameter distribution includes as special cases models (4), (5), and (6) above, but when the sample resembles a Poisson distribution the models tend to coincide. When the variance is less than the mean no fit is possible. When zeros are missing no fit is attempted.

The parameters chosen for fitting the distribution are functions of the expected moments as follows:

$$\theta_1 = m = \lambda \pi k,$$

$$\theta_2 = \frac{v - m}{m^2} = \frac{k + 1}{\lambda k},$$

$$\theta_3 = \frac{m(m_3 - 3v + 2m)}{(vm)^2} = \frac{k + 2}{k + 1},$$

from which λ, π, and k may be derived. If $\theta_2 = 0$ the distribution is Poisson regardless of θ_3, which fact distorts the likelihood contours in the neighborhood of $\theta_2 = 0$. The model is overparamtrized for such data, because the parameters θ_2 and θ_3 characterize the underlying distribution of the means of the individual Poisson, for which no information is available. For larger values of θ_2 the model is easily fitted, with θ_3 generally lying between 1 (corresponding to the Neyman Type A) to 2 (corresponding to the negative binomial).

7.3.2. Fitting Continuous Distributions

The continuous distributions fitted by MLP may be considered in groups. The normal distribution, and mixtures of two normals assume data with unrestricted range. The lognormal, exponential, Weibull, and gamma distributions assume positive observations. Two related forms of the beta distribution are available: the beta type 1 is for observations in the range (0, 1) and the beta type 2 is for positive observations.

The choice of stable parameters depends on the underlying shape of the distribution. For normal-like samples mean and variance are stable, but for long-tailed distributions medians and other percentiles are more likely to be stable.

When there are missing observations (censored classes) it is assumed that stability is not seriously affected. The likelihood is based on the observed data, and the estimated frequencies in the missing cells are computed from the estimated total sample size, $\hat{N} = N/(1 - \sum p_i)$, where the p_i are the calculated probabilities for the cells which are not missing.

1. *Mixtures of two normal distributions*

The most general model has five parameters, two means (μ_1 and μ_2), two variances (σ_1^2 and σ_2^2), and the proportion in the first component ($0 < \pi < 1$). The expected mean and variance of the mixed distribution are

$$m = \pi\mu_1 + (1 - \pi)\mu_2,$$
$$v = \pi\sigma_1^2 + (1 - \pi)\sigma_2^2 + \pi(1 - \pi)(\mu_1 - \mu_2)^2.$$

On the assumption that m and v are suitable stable parameters, and that σ_1^2 and σ_2^2 cannot exceed v, it is convenient to work with the parameter set $(m, v, \pi, \sigma_1, \sigma_2)$. To obtain initial estimates for π, σ_1, and σ_2, parameters m and v are at first constrained to equal the sample mean and variance. The full five-parameter model is then fitted, with little danger of converging to unwanted local solutions associated with extreme values of π and σ_1 or σ_2. Lower limits are placed on σ_1 and σ_2 to exclude the solutions which correspond to a single class of the data being fitted by one component with very small variance, while the remaining classes are fitted by the overall normal distribution. Limits on π of (0.02, 0.98) help to exclude other extreme cases.

If there is no evidence for two components, or if the means μ_1 and μ_2 are very close, π becomes indeterminate and the solution depends critically on the difference in the variances. Two equivalent solutions always exist, obtained by exchanging suffixes 1 and 2 and replacing π by $(1 - \pi)$. The program may find either solution, although a convention could be adopted that $\mu_1 \le \mu_2$.

Two special cases are provided which are fitted in a similar way. It is often simpler to assume $\sigma_1^2 = \sigma_2^2$, and usually this makes very little different to the fit. In some theoretical situations y is known, and the program provides the solution for $\pi = 0.5$, σ_1^2. The single normal may also be fitted, but this presents no difficulty.

Mixtures of lognormals are not programmed separately as they may be fitted by transforming the data.

2. *Lognormal with unknown origin*

The three-parameter lognormal with unknown origin (which must be strictly less than the minimum data observation), can be fitted by a one-parameter search for the origin parameter. Since $y = \log(x - \alpha)$ is assumed to be normal, the sample mean and variance of y are substituted for the theoretical parameters μ and σ^2 in the log-likelihood function. If the sample is negatively skewed before transformation, no value of α can be fitted, and the limiting case, $\alpha = -\infty$, corresponds to a normal distribution. At the other end of the scale there are data sets for which $\hat{\alpha}$ is the minimum value of x, and when the individual observations are used to evaluate the log-likelihood there is always a local optimum at this point. The search range for α is therefore limited to a small quantity less than the first class interval.

Initial estimates α_0 of α are obtained from three estimated sample percentiles x_{25}, x_{50}, and x_{75} as follows:

$$\alpha_0 = \frac{x_{25}x_{75} - x_{50}^2}{x_{25} + x_{75} - 2x_{50}},$$

from which convergence is usually rapid unless either of the extreme cases are encountered. The three-parameter optimization with respect to α, μ, and σ completes the fit.

3. *Weibull distribution with known origin*
The two-parameter Weibull distribution

$$F(x) = 1 - \exp(-bx^c)$$

(where $c = 1$ corresponds to an exponential distribution) is easily fitted using the median, x_{50}, as a stable parameter, along with c (which is usually in the range 0.1–5). Given c and x_{50}, the parameter b is found to be $(\log 2)^{1/c}/x_{50}$. Empirically, x_{50} is less correlated with c, but the median is simple to estimate and the fit is usually obtained with no difficulty.

The three-parameter Weibull with unknown origin presents problems when c is less than one, because the Weibull distribution has an infinite mode at $x = 0$, and the origin parameter is always estimated by the minimum value of x in the sample. This form is not provided by MLP directly.

4. *Gamma distribution with known origin*
The two-parameter gamma distribution with frequency function

$$f(x) = \frac{b(bx)^{k-1}e^{-bx}}{\Gamma(k)}$$

corresponds to the exponential distribution when $k = 1$. For practical purposes it is not very different from the Weibull distribution, and is fitted in similar manner in terms of the median, x_{50}, and the reciprocal of the parameter, k. Given k, the median of the standard gamma function (with $b = 1$) is evaluated using the Cornish–Fisher expansion of the distribution. Using the observed median x_{50} gives a method for calculating b. The initial estimate of k is obtained from m^2/v using the sample mean and variance; a direct optimization with respect to b and k is then performed.

The gamma distribution with unknown origin presents similar problems to the Weibull, and is not provided explicitly.

5. *Beta distributions of the first kind*
The beta distribution of the first kind

$$F(x) = \frac{x^{p-1}(1 - x)^{q-1}}{B(p, q)},$$

applies only to x in the range $(0, 1)$. The model is easily fitted in terms of the reciprocals of parameters p and q, obtaining initial estimates from the sample mean and variance using the formulas

$$\frac{1}{p_0} = \left(\frac{m(1 - m)}{v} - 1\right)m, \qquad \frac{1}{q_0} = \left(\frac{m(1 - m)}{v} - 1\right)(1 - m).$$

The three-parameter beta distribution of the second kind,

$$F(x) = \frac{1}{Bp_1 q}\frac{b^p x^{p-1}}{(1 + bx)^{p+q}},$$

applies to positive x, and includes the scale parameter b which has the same role as in the gamma and Weibull distributions. An approximate formula for the median, given p and q, is used, along with the sample median, to estimate b, given trial values of $1/p$ and $1/q$. Finally, the three parameters b, p, and q are fitted directly.

The extreme cases in which either $1/p$ or $1/q$ are 0 correspond to the gamma distribution. More highly skewed samples cannot be fitted at all, so that if no optimum is found with positive parameters the limiting case (gamma) is fitted.

7.4. Standard Biological Models Requiring Maximum Likelihood Estimation

Since MLP was written to serve the needs of biological scientists, models appropriate to standard laboratory practices are included. Many of these have wider application than was originally envisaged; for example, the probit regression method has been used to measure the breaking stresses of building materials, in addition to its traditional use in toxicity testing. The underlying statistical analyses are all of the class: nonlinear expectation, nonnormal error distribution; some are generalized linear models (McCullagh and Nelder, 1982). The output given by MLP is in terms of the practical needs of the laboratory scientists.

7.4.1. Estimating Bacterial Density by the Dilution Series Technique

The dilution series technique of estimating bacterial density relies on the assumption that if one or more bacteria are present in any subsample of the original volume, their presence may be detected by their activity in a suitable growth medium. Samples showing no activity are assumed to be sterile (containing no bacteria). The original volume, assumed to contain an unknown number of bacteria, λ, to be estimated, is diluted by the addition of $\alpha - 1$ equal volumes of sterile material, and is throughly mixed and subdivided into equal volumes assumed to include a fraction $1/\alpha$ of the original volume. Some of these subsamples may be tested for sterility, and the remainder further diluted until a high proportion are expected to be sterile.

If n_i samples are tested at the ith dilution, the number of fertile (nonsterile) samples is r_i, which is distributed binomially with expected proportion

$$1 - \exp\left(-\frac{\lambda}{\alpha^i}\right),$$

on the assumption of completely random allocation to subsamples, using the Poisson distribution. If the allocation is nonrandom, as may occur if it is difficult to ensure that all clumps of adjacent bacteria are separated, a negative binomial distribution may be more realistic, and the expected proportion is then

$$1 - \left(1 + \frac{\lambda}{k\alpha^i}\right)^{-k},$$

where k is the negative binomial parameter (see Section 7.3.1(4)).

Fisher (as quoted in Fisher and Yates, 1963, Table VIII$_2$), found that $\log \lambda$ is an approximately stable parameter, and in balanced designs with equal n_i at each dilution the estimate depends mainly on the mean fertile level, $\sum r_i / \sum n_i$. An explanation for the stability of $\log \lambda$ is that if the expected response at some median value j is well estimated, then equal changes in j imply proportional changes in λ if λ/α^j is to remain constant.

MLP uses the mean fertile level to obtain an initial estimate of $\log \lambda$, except in the two extreme cases where the samples are all fertile or all sterile, in which the MLEs of λ are infinity and zero, respectively. The program finds exact confidence limits for $\log \lambda$ (and λ) in each case; when the samples are all fertile or all sterile the appropriate one-sided limit is given.

If the fit is poor for the Poisson model, the negative binomial model may be used instead. The program starts from the Poisson fit and fits the parameter $1/k$ to account for the greater spread of responses. The estimate of $\log \lambda$ is reduced as $1/k$ increases, and its confidence limits are considerably widened. The negative binomial model is therefore only used where it is reasonable to assume lack of randomness in the mixing process.

7.4.2. Probit Analysis and its Variants

The term *probit analysis* (Finney, 1974) derives from the early use of the cumulative normal distribution (the probability integral) to transform proportions so that response curves become linear. Its traditional area of application is in modeling the increase in the proportion of test subjects responding as a stimulus is increased. Examples are the proportion of insects killed by a given dose of insecticide, the proportion of subjects in s psychology laboratory test who correctly identify the stronger of two given signals, or the proportion of children of given age who pass a particular academic test.

The standard model, in which groups of n subjects are tested for a range of values of a stimulus x (often a log concentration, in which zero is treated separately) and r subjects respond in a particular way, assumes that the expected probability, $p = f(\alpha + \beta x)$, is an increasing function of x in the interval $(0, 1)$. The inverse transformation, $f^{-1}(p) = y$, may be plotted against x to show the linear relationship, and the fitted parameter β is known as the *slope* of the response (after transformation).

The parameters of interest are the LD_{50} (median lethal dose), which is the value of x corresponding to $p = 0.5$, and the slope. When the aim of the study is to compare one stimulus with another, such as a new insecticide compared with a standard, a great simplification is to constrain the slopes to be equal (parallel line analysis), and to estimate the difference between the LD_{50} values for each line and first line, which in toxicity studies gives the *relative potency* on a log scale.

The main variants of the standard analysis are required to meet practical difficulties in conducting the ideal tests.

1. In some circumstances the responses do not lie between 0 and 1 but between c_1 and $1 - c_2$, where c_1 is known as the *control mortality* and c_2 as the *high-dose immunity*. This nomenclature refers to toxicity tests in which a proportion c_1 are affected by factors independent of the stimulus x, and a proportion c_2 are unaffected by any level of x. The expected response to x then is modified to

$$p^* = c_1 + (1 - c_1 - c_2)p.$$

2. When the number tested at each level of x is unknown as, for example, when eggs of insects concealed in grains of corn are treated with insecticide, and only those that survive to emerge are counted, the probability distribution is Poisson rather than binomial, and the mean value of n must be treated as an extra paramter to be estimated. This is known as *Wadley's problem* (Finney, 1974).

3. When two independent stimuli x_1 and x_2 are applied simultaneously the simplest model is that

$$p = f(\alpha + \beta_1 x_1 + \beta_2 x_2),$$

which is known as the *probit plane*. More complicated models have been suggested to allow for interaction between x_1 and x_2, but these are not provided as standard options.

4. In the study of resistance to insecticides it is found that the response may become decidedly nonlinear (on the transformed scale) if a mixture of resistant and susceptible insects are tested simultaneously. Although there is usually not enough information from a single set of responses of the form

$$p^* = \pi f(\alpha_1 + \beta x) + (1 - \pi)f(\alpha_2 + \beta x)$$

to estimate the extra parameters π and α_2, there may be additional tests performed on pure resistant and pure susceptible populations which then enable π to be estimated.

5. In tests of herbicides on growing plants the effects may be expressed as a proportional diminution of growth relative to untreated control plants. In this case the response is quantitative, and a lognormal error mdoel is assumed (modified to allow for plants that are killed completely). Although this model can also be considered purely in terms of curve fitting, it is important for the users to be able to interpret it as an assay, so that the effects of the herbicide and the residues being measured, be estimated directly.

The options provided by MLP under the heading probit analysis are as follows:

(1) *Three linearizing transformations*:
   (a) *Probit* or normal equivalent deviate, in which $f(x)$ is the normal probability integral.
   (b) *Logit*, where

$$y = \tfrac{1}{2} \log\left(\frac{p}{1-p}\right),$$

   or

$$f(x) = \frac{1}{1 + \exp(-2x)},$$

   where the factor 2 is introduced for comparability with the probit transformation. Many writers omit the factor 2, which make no difference to the fit or to the conclusions.
   (c) *Complementary log log*, where

$$y = \log(-\log(1-p)),$$

   or

$$f(x) = 1 - \exp(-\exp(x)),$$

   which is derived from a model which depends on the proportion of zeros in the Poisson distribution, a speical case of which is used in the dilution series analysis of Section 7.4.1.
(2) The control mortality c_1 may be input as a constant, or estimated from the data. Samples corresponding to the control concentration ($x = -\infty$) are indicated. The high-dose immunity may also be supplied or estimated, but it is not possible to include a data value for $x = +\infty$.
(3) Data may be appropriate for the ordinary binomial probit line, for Wadley's problem (survivors only), for the probit plane, or for the analysis of mixtures of two populations.
(4) Parallel line assays are available for the probit line model, with analysis of parallelism to test for differences of slope and of position, given equal slopes.

Output includes LD_{50} and other percentage points, where the standard errors are given only if the slope is significantly different from zero. Approximate fiducial limits based on Fieller's theorem are given, although in very small samples these may be rather different from the exact likelihood-based confidence intervals (see Fig. 3.10).

Standard errors of parameter estimates may be increased by a heterogeneity factor if the residual deviance is greater than the degrees of freedom. This is appropriate where it has not been possible to guarantee random allocation of subjects to dose levels, and the lack of fit is not ascribable to systematic departures from the model.

The parameters used in fitting the models are the coefficients of linear functions of $(x - x^*)$ where x^* is a provisional weighted mean of x using weights based on the observed proportions. During the preliminary phase, data sets for which the MLE of the slope is infinite are excluded, and a diagnostic message is produced. To obtain an estimate of slope at least two responses must lie between the lower and upper limits.

When the control mortality parameter c_1 is estimated from data rather than fixed, its estimate may be negative when the lowest observations indicate a steeper response than the theoretical curve; if the data include a control group with nonzero response this situation cannot arise. Similar problems may occur if c_2 is estimated in addition.

In Wadley's problem the mean number of subjects in each group is a parameter to be estimated, which is easily achieved by treating it as a separable linear parameter (Section 7.1.1(3a)). The fitted slope may be infinite if the number of survivors changes with increasing dose, from a level average response, to zero.

7.4.3. Quantitative Assays

Although many models for quantitative assays may be fitted by parallel curve analysis (Section 7.2.2(4)) a special section of MLP is available for the particular case where the first set of data is a standard logistic response curve, and subsequent sets refer to samples where the absolute values of x are unknown, their relative values being the dilution levels derived from the experimental technique.

For these data the lognormal error distribution is assumed, and the parametrization is based on the idea that the expectation at the mean value of x in each sample is likely to be a stable parameter. The first set of data, which usually contains more observations, is fitted as described in Section 7.2.8 for the logistic curve.

7.5. General User-Defined Models in MLP

Details of the facilities for fitting user-defined models will not be described here. The whole of this book has depended on the ability to define models in various forms and to test their validity for different data sets.

Very briefly, the facilities consist of procedures for:

(a) defining the model in terms of the parameters, using high-level interpretive instructions allowing mathematical functions, discrete operations, solution of differential equations, integration, and operating on successive values;

(b) selection of error distribution (including choice of weighting);

(c) data description, including parallel subsets;

(d) control of optimization (choice of algorithm, initial estimates, steps, and bounds);

(e) tabular output (fitted values, first derivatives, parameter loadings, standard errors of prediction);

(f) graphical output (contour plots, graphs of fitted functions);

(g) confidence regions (limits on parameters and functions);

(h) parallel model analysis (common parameters, common nonlinear parameters, displacements in one variable); and

(i) diagnostics (the "hat" matrix, variance multipliers).

Details of these facilities will be found in the MLP manual (Ross, 1987).

MLP also contains facilities for estimating parameters in special models in genetics, in transition matrix models for multivariate counts at discrete intervals of time, and for general linear regression. Details of these models may also be found in the MLP manual.

Glossary of Unfamiliar Terms Used in this Work

A posteriori stable parameters
Parameters whose stability can only be established after the model has been fitted in terms of other parameters (Section 2.3).

A priori stable parameters
Parameters for which stability is expected on informal grounds before the model is fitted.

Analogous parameters
Parameters in different models which have the same practical interpretation, such as asymptotes of curves (Section 2.5).

Analysis of deviance
Generalization of analysis of variance which uses the deviances obtained by fitting hierarchically related models with different numbers of parameters (Section 1.2.1).

Calibration model
A type of parallel model in which the same response function is expected for each set of data, but the position in each set relative to the first is to be estimated (Section 1.1.2).

Common estimate locus
The locus in data space of all data sets for which a given parameter value is a maximum likelihood estimate (MLE) (Section 4.3.1).

Computing parameters
Parameters chosen for efficient computation of maximum likelihood estimates (MLEs) (Section 2.1).

Convergence locus
The locus in parameter space of all initial estimates which converge to the solution in a given number of iterations, given the estimation procedure and the convergence criterion (Section 4.4.8).

Defining parameters
Parameters that define a model in standard algebraic terms (Section 2.1).

Design (nonlinear)
Controllable choice of observations, such as timing and frequency of making a measurement (Section 3.3).

Deviance
Measure of discrepancy between data and model based on the likelihood ratio test statistic (Section 1.2.1).

Deviance residual
Residual computed from the square root of the contribution to the deviance attributed to any given observation, taking the same sign as the simple residual (Section 2.4).

Discrepancy plot
Plot of difference between a general function and an approximation to the function (Section 4.4.2).

Effective design matrix
Matrix of first derivatives of expectations with respect to parameters, being the analogue of the design matrix in a linear model (Section 3.3.1).

Effective replication
The number of observations (not necessarily an integer) required to estimate a particular expected value with the same precision as the fitted value from the given model and data (Section 3.3).

Function loadings
The relative effects of each observation in a data set in estimating a function of parameters (Section 4.2).

Function locus
The locus in data space of functions of parameters: a subspace of the solution locus (q.v.) (Section 4.3.1).

Hat matrix
In a linear model, the matrix H such that $\hat{y} = Hy$, the matrix that puts on the "hat." In a nonlinear model, an approximation may be computed using the effective design matrix (q.v.) (Section 3.3).

Interpretable parameters
Parameters that relate directly to quantities of interest in the model and the data (Section 2.1).

Intrinsic nonlinearity
Nonlinearity in the model that cannot be removed by transforming parameters, explained in terms of curvature of the solution locus (Section 2.4.4).

Likelihood-based confidence regions
Regions of parameter space bounded by contours of equal likelihood. Given a particular contour, corresponding to a likelihood ratio test level of significance, upper and lower limits for particular parameters or functions of parameters may be found numerically (Section 1.2.2).

Linear parameters
Parameters β_k in models of the form $E(y) = \sum \beta_k f_k$ where f_k may be a constant, an independent variable, or a function of nonlinear parameters. When the error distribution is normal, Poisson, or gamma some linear parameters may be estimated directly in terms of functions of nonlinear parameters. Such parameters are called *separable* (Section 4.4.3).

Parallel models
By analogy with parallel lines, models fitted to several sets of data with some parameters common to all sets, the rest specific for each set (Section 1.1.2).

Parameter-effects curvature
Nonlinearity due to choice of parameter system, additional to intrinsic nonlinearity (q.v.) (Section 2.4.4.)

Parameter loadings
The relative effects of particular observations on the estimates of each parameter (Section 3.3.2).

Parameter locus
The locus in parameter space of all parameter sets predicting a given fitted value (Section 4.4.5).

Pseudomodel
A model introduced to extend the range of expectations so that data may be fitted and likelihood contours evaluated. Parameters in pseudomodels may not be interpretable in the terms of the original model, but in combination with estimates from other data sets may lead to acceptable inferences (Section 3.1.4).

Self-estimating observations
Observations with effective replication close to unity, so that their expectations derive almost no information from other observations via the model (Section 3.3.1).

Separable linear parameters, See *Linear parameters*

Similar models
Models expressible in terms of *analogous parameters* (q.v.), such as curves of the same general shape over a range of data values (Section 2.5).

Solution locus
The locus in data space of expectations generated by different values of the parameters: the set of data values exactly fitted by the model. Also called the *expectation surface* (Section 4.3.1).

Stable ordinates
In curve fitting, sets of points at which the expectations are uncorrelated, or nearly so (Section 2.3).

Stable parameters
Parameters not greatly affected by uncertainties in the values of other parameters (Section 2.1.3).

Statistic locus
In data space the locus of all data values sharing the same value of some statistic such as the mean (Section 4.3.2).

Working constant
A constant estimated *a priori* to relate a general class of transformations to the particular locations, scales or other properties of data sets (Section 2.2).

References*

AITCHISON, J. and BROWN, J.A.C. (1957). *The Lognormal Distribution*. Cambridge: Cambridge University Press.

ANDREWS, D.F., BICKEL, P.J., HAMPEL, F.R., HUBER, P.J., ROGERS, W.H., and TUKEY, J.W. (1972). *Robust Estimates of Location*. Engelwood Cliffs, NJ: Princeton University Press.

ANSCOMBE, F.J. (1950). Sampling theory of the negative binomial and logarithmic series distribution. *Biometrika* **37**, 358–382.

ATKINSON, A.C. (1982). Regression diagnostics, transformations and constructed variables (with discussion). *Journal of the Royal Statistical Society B* **44**, 1–36.

BARD, Y. (1974). *Nonlinear Parameter Estimation*. New York: Academic Press.

BARHAM, R.H. and DRANE, W. (1972). An algorithm for least squares estimation of non-linear parameters when some parameters are linear. *Technometrics* **14**, 757–766.

BARNETT, V.D. and LEWIS, T. (1984). *Outliers in Statistical Data* (2nd ed.). New York: Wiley.

BATES, D.M. and WATTS, D.G. (1980). Relative curvature measures of non-linearity (with discussion). *Journal of the Royal Statistical Society B* **42**, 1–25.

BATES, D.M. and WATTS, D.G. (1981). Parameter transformations for improved approximate confidence regions in non-linear least squares. *Annals of Statistics* **9**, 1152–1167.

BEALE, E.M.L. (1960). Confidence regions on non-linear estimation (with discussion). *Journal of the Royal Statistical Society B* **22**, 41–88.

BEALE, E.M.L., KENDALL, M.G., and MANN, D.W. (1967). The discarding of variables in multivariate analysis. *Biometrika* **54**, 357–365.

BEAUCHAMP, J. and CORNELL, R.G. (1966). Simultaneous non-linear estimation. *Technometrics* **8**, 319–326.

BELSLEY, D.A., KUH, E., and WELSCH, R.E. (1980). *Regression Diagnostics*. New York: Wiley

BERZUINI, C., ROSS, G.J.S., and LARIZZA, C. (1986). Developing intelligent software for

---

* Including some relevant references not referred to in the text.

nonlinear model fitting as an expect system. In: *COMPSTAT 1986*. Heidelberg: Physica-Verlag, pp. 259–264

BLISS, C.I. (1935). The calculation of the dosage–mortality curve. *Annals of Applied Biology* 22, 134–167.

BLISS, C.I. and JAMES, A.T. (1966). Fitting the rectangular hyperbola. *Biometrics* 22, 573–602.

BOX, G.E.P. and COX, D.R. (1964). An analysis of transformations. *Journal of the Royal Statistical Society B* 26, 211–252.

BOX, G.E.P. and HUNTER, W.G. (1962). A useful method of model building. *Technometrics* 4, 301–318.

BOX, G.E.P. and JENKINS, G.M. (1970). *Time Series Analysis, Forecasting and Control.* San Francisco: Holden Day.

BOX, G.E.P. and LUCAS, H.L. (1959). Design of experiments in non-linear situations. *Biometrika* 46, 77–90.

BOX, M.J. (1968). The occurrence of replications in optimal designs of experiments to estimate parameters in non-linear models. *Journal of the Royal Statistical Society B* 30, 290–302.

BOX, M.J. (1971). Bias in non-linear estimation. *Journal of the Royal Statistical Society B* 32, 171–201.

BROOKS, R.S., DAWID, A.P., GALBRAITH, J.I., GALBRAITH, R.F., STONE, M., and SMITH, A.F.M. (1978). A note on forecasting car ownership. *Journal of the Royal Statistical Society A* 141, 64–68.

BROWNIE, C. and ROBSON, D.S. (1976). Models allowing for age-dependent survival rates for band return data. *Biometrics* 32, 305–323.

BUNKE, H. (1980) Parameter estimation in non-linear regression models. In: *Handbook of Statistics*, Vol. 1. Ed. P.R. Krishnaiah. Amsterdam: North-Holland, Elsevier, pp. 593–615.

BUWALDA, J.G., ROSS, G.J.S., STRIBLEY, D.B., and TINKER, P.B. (1982). The development of endomycorrhizal root systems. IV. The mathematical analysis of the effect of phosporus on the spread of vesicular–arbuscular mycorrhizal infection in root systems. *New Phytologist* 92, 391–399.

CAUSTON, D.R. and VENUS, J.C. (1981). *The Biometry of Plant Growth.* London: Arnold.

CHAMBERS, J.R. (1973). Fitting non-linear models: numerical techniques. *Biometrika* 60, 1–13.

CHIEN-FU, W.U. (1981). Asymptotic theory of non-linear least squares estimation. *Annals of Statistics* 9, 501–513.

CORNELL, R.G. (1962). A method of fitting linear combinations of exponentials. *Biometrics* 18, 104–113.

COX, D.R. (1962). Further results on tests of separate families of hypotheses. *Journal of the Royal Statistical Society B* 24, 406–424.

COX, D.R. and HINKLEY, D.V. (1974). *Theoretical Statistics.* London: Chapman & Hall.

CRAMER, H. (1946). *Mathematical Methods of Statistics.* Engelwood Cliffs, NJ: Princeton University Press.

DANIEL, C. and WOOD, F.S. (1973). *Fitting Equations to Data.* New York: Wiley–Interscience.

DAVIDON, W.C. (1959). Variable metric method for minimization. Report ANL-5590, Argonne Natural Library.

DAY, N.E. (1969). Estimating the components of a mixture of normal distributions. *Biometrika* 56, 463–474.

DEMPSTER, A.P., LAIRD, N.M. and RUBIN, D.B. (1977). Maximum likelihood from incomplete data via the EM algorithm (with discussion). *Journal of the Royal Statistical Society B* 39, 1–38.

DIXON, L.C.W. (1972). *Non-linear Optimisation*. London: English Universities Press.
DRAPER, N.R. and SMITH, H. (1966, 1981). *Applied Regression Analysis*. New York: Wiley.

EDWARDS, A.W.F. (1972). *Likelihood*. Cambridge: Cambridge University Press.

FEDOROV, V.V. (1972). *Theory of Optimal Experiments*. Trs. W.J. Studden & E.M. Klimko. New York: Academic Press.
FIELLER, E.C. (1954). Some problems in interval estimation. *Journal of the Royal Statistical Society B* **16**, 175–185.
FINNEY, D.J. (1947, 1952). *Probit Analysis*. Cambridge: Cambridge University Press.
FINNEY, D.J. (1978). *Statistical Method in Biological Assay* (3rd ed.). London: Griffin.
FISHER, R.A. (1939). The sampling distribution of some statistics obtained from non-linear equations. *Annals of Eugenics* **9**, 238–249.
FISHER, R.A. (1956). *Statistical Methods and Scientific Inference*. Edinburgh: Oliver & Boyd.
FISHER, R.A. and YATES, F. (1974). *Statistical Tables for Biological Agricultural and Medical Research* (6th ed). London: Longman.
FORSYTHE, G.E. (1957). Generation and use of orthogonal polynomials for data fitting with a digital computer. *Journal of the Society Industrial Applied Mathematics* **5**, 74–88.

GART, J.J. (1970). *Random Counts in Scientific Works* **1**, 171–191.
GOMES, P.F. (1953). The use of Mitscherlich's regression law in the analysis of experiments with fertilisers. *Biometrics* **9**, 498–516.
GOULDEN, C.H. (1939). *Methods of Statistical Analysis*. New York: Wiley.
GREEN, P.J. (1984). Iteratively reweighted least squares for maximum likelihood estimation and some robust and resistant alternatives (with discussion). *Journal of the Royal Statistical Society B* **46**, 149–192.
GROSENBACH, L. (1965). Generalisation and reparametrisation of some sigmoid or other non-linear functions. *Biometrics* **21**, 708–714.
GUTTMAN, I. and MEETER, D.A. (1965). On Beale's measure of non-linearity. *Technometrics* **7**, 623–637.

HALPERIN, M. (1963). Confidence interval estimation in non-linear regression. *Journal of the Royal Statistical Society B* **63**, 330–333.
HALPERIN, M. and MANTEL, N. (1963). Interval estimation of non-linear parametric functions. *Journal of the American Statistical Association* **58**, 611–627; **59**, 168–181; **60**, 1191–1199.
HARTLEY, H.O. (1948). The estimation of non-linear parameters by "internal least squares." *Biometrika* **35**, 32–45.
HARTLEY, H.O. (1961) The modified Gauss–Newton method for the fitting of non-linear regression functions by least squares. *Technometrics* **3**, 269–280.
HARTLEY, H.O. (1964). Exact confidence regions for the parameters in non-linear regression laws. *Biometrika* **51**, 347–354.
HARTLEY, H.O. and BOOKER, A. (1965). Non-linear least squares estimation. *Annals of Mathematical Statistics* **36**, 638–650.
HAWKINS, D.M. (1980). *Identification of Outliers*. London: Chapman & Hall.
HOERL, A.E. and KENNARD, R.W. (1970). Ridge regression: biased estimation for non-orthogonal problems. *Technometrics* **12**, 55–67.

JENNRICH, R.I. (1969). Asymptotic properties of non-linear least squares estimation. *Annals of Mathematical Statistics* **40**, 633–643.
JENNRICH, R.I. and MOORE, R.J. (1975). Maximum likelihood estimation by means of

non-linear least squares. In: *Proceedings of the American Statistical Association (Statistical Computing Section)*, 57–65.

JOHNSON, N.L. and KOTZ, S. (1969). *Distributions in Statistics* (four volumes). New York: Wiley.

JØRGENSON, B. (1983). Maximum likelihood estimation and large sample inference for generalised linear and non-linear regression models. *Biometrika* **70**, 19–28.

KALBFLEISCH J.D. and SPROTT, D.A. (1970). Application of likelihood methods to models involving large numbers of parameters. *Journal of the Royal Statistical Society B* **32**, 175–208.

KENDALL, M.G. and STUART, A. (1967). *Advanced Statistics* (Vol. 2). London: Griffn.

KOSHAL, R.S. (1933) Application of the method of maximum likelihood to the improvement of curves fitted by the method of moments. *Journal of the Royal Statistical Society* **96**, 303–313.

KOTZ, S. and JOHNSON, N. (eds). (1985). *Encyclopedia of Statistical Sciences* (Nonlinear models, nonlinear regression.) New York: Wiley.

LAWTON, W.H. and SYLVESTRE, E.A. (1971). Elimination of linear parameters in non-linear regression. *Technometrics* **13**, 461–467.

LIPTON, S. and MCGILCHRIST, C.A. (1963). Maximum likelihood estimation of parameters in double exponential regression. *Biometrics* **19**, 144–151.

LIPTON, S. and MCGILCHRIST, C.A. (1964). The derivation of methods for fitting exponential regression curves. *Biometrika* **51**, 504–507.

MCCULLAGH, P. and NELDER, J.A. (1983). *Generalized Linear Models*. London: Chapman & Hall.

MALINVAUD, E. (1970). The consistency of non-linear regression. *Annals of Mathematical Statistics* **41**, 956–969.

MICHAELIS, L. and MENTEN, M.L. (1913). Kinetik der Invertinwirkung. *Biochemische Zeitschrift* **49**, 333.

NAMKOONG, G. and MILLER, D.L. (1968). Estimation of non-linear parameters for non-asymptotic function. *Biometrics* **24**, 439–440.

NELDER, J.A. (1961). The fitting of a generalization of the logistic curve. *Biometrics* **17**, 89–110.

NELDER, J.A. (1962). An alternative form of a generalized logistic equation. *Biometrics* **18**, 614–616.

NELDER, J.A. (1966). Inverse polynomials, a useful group of multifactor response functions. *Biometrics* **22**, 128–141.

NELDER, J.A. and WEDDERBURN, R.W.M. (1972). Generalised linear models. *Journal of the Royal Statistical Society A* **135**, 370–384.

NEYMAN, J. and PEARSON, E.S. (1928). On the use and interpretation of certain test criteria for the purposes of statistical inference. *Biometrika* **20a**, 175–240 and 263–294.

PATTERSON, H.D. (1956). A simple method for fitting an asymptotic regression curve. *Biometrics* **12**, 323–329.

PATTERSON, H.D. (1969). Baule's equation. *Biometrics* **29**, 159–164.

PEARSON, K. (1894). Contributions to the mathematical theory of evolution. *Phil. Trans. Royal Society A* **185**, 71–110.

PEARSON, K. and MOUL, M. (1925). *Annals of Eugenics* **1**, 41.

RALSTON, M.L. and JENNRICH, R.I. (1978). DUD, a derivative free algorithm for non-linear least squares. *Technometrics* **20**, 7–14.

RATKOWSKY, D.A. (1984). *Non-linear Regression Modeling*. New York: Dekker.

RICHARDS, F.J. (1959). A flexible growth curve for empirical use. *Journal of Experimental Botany* **10**, 290–300.

RICHARDS, F.S.G. (1961). A method of maximum likelihood estimation. *Journal of the Royal Statistical Society B* **23**, 469–476.

Ross, G.J.S. (1970). The efficient use of function minimization in non-linear maximum likelihood estimation. *Applied Statistics* **19**, 205–221.

Ross, G.J.S. (1975). Simple non-linear modelling for the general user. In: *Proceedings of the 40th Session. International Statistical Institute, Warsaw* **2**, 585–593.

Ross, G.J.S. (1978). Exact and approximate confidence regions for functions of parameters in non-linear models. In: *COMPSTAT 1978*. Wien: Physica-Verlag, pp. 110–116.

Ross, G.J.S. (1980). Uses of non-linear transformations in non-linear optimisation problems. In: *COMPSTAT 1980*. Wien: Physica-Verlag, pp. 381–388.

Ross, G.J.S. (1982). Least squares optimisation of general log-likelihood functions and estimation of separable linear parameters. In: *COMPSTAT 1982*. Wien: Physica-Verlag, pp. 406–411.

Ross, G.J.S. (1984). Parallel model analysis: fitting non-linear models to several sets of data. In: *COMPSTAT 1984*. Wien: Physica-Verlag, pp. 458–463.

Ross, G.J.S. and PREECE, D.A. (1985). The negative binomial distribution. *The Statistician* **34**, 323–336.

Ross, G.J.S. (1987). *Maximum Likelihood Program*. Numerical Algorithms Group, Oxford.

RUTHERFORD, E. and GEIGER, H. (1910). The probability variations in the distributions of alpha particles. *Philisophical Magazine*, 6th ser., **20**, 698–704.

SHAH, B.K. (1961). A simple method of fitting the regression curve $y = \alpha + \delta x + \beta \rho^x$. *Biometrics* **17**, 651–652.

SNEDECOR, G.W. (1946). *Statistical Methods*. 4th ed. Iowa State College Press.

SOLARI, M.E. (1969). The "maximum likelihood solution" to the problem of estimating a linear functional relation. *Journal of the Royal Statistical Society B* **31**, 372–375.

SPRENT, P. (1969). *Models in Regression and Related Topics*. London: Methuen.

STEVENS, W.L. (1951). Asymptotic regression. *Biometrics* **7**, 247–267.

TOOTILL, J.P.R. (1963). Routine least squares estimation from models containing a single non-linearity. *Biometrics* **19**, 118–143.

TORNHEIM, T. (1963). Convergence in non-linear regression. *Technometrics* **5**, 513–514.

TURNER, M.E., MONROE, R.J., and LUCAS, H.L. (1961). Generalised asymptotic regression and non-linear path analysis. *Biometrics* **17**, 120–143.

VILLEGAS, C. (1969). On the least squares estimation of non-linear relations. *Annuals of Mathematical Statistics* **39**, 467–474.

WAGNER, H.M. (1962). Non-linear regression with minimal assumptions. *Journal of the American Statistical Association* **57**, 572–578.

WALSH, G.R. (1975). *Methods of Optimisation*. New York: Wiley.

WILLIAMS, E.J. (1962). Exact fiducial limits in non-linear estimation. *Journal of the Royal Statistical Society B* **24**, 125–139.

WILLSON, L.J., FOLKS, J.L., and YOUNG, J.H. (1984). Multiple estimation compared with fixed sample size estimation of the negative binomial parameter k. *Biometrics* **40**, 109–118.

ZANAKIS, S.H. and RUSTAGI, J.S. (1982). *Optimization in Statistics*. Amsterdam: North-Holland, Elsevier.

Author Index

Subject Index